HARCOU

Math

South Carolina PACT Success!

Grade 4

Harcourt

Orlando • Boston • Dallas • Chicago • San Diego
www.harcourtschool.com

Photo Credits: iii(l) & 7, Robin Nelson; iii(r) & 17, Jane Faircloth/Transparencies Inc.; iv & 41, 25, & 50, Michael Meekins; 3, Superstock; 11 & 45, Larry Gleason; 29, Gary Knight; 30, Columbia Museum of Art; 36, Billy E. Barnes/Photo Edit.

Printed in the United States of America

ISBN   0-15-325894-2

1 2 3 4 5 6 7 8 9 10 129 10 09 08 07 06 05 04 03 02 01

© Harcourt

Introduction

South Carolina PACT Success! develops student confidence in taking the PACT, connects math to real life in South Carolina, and provides practice and review of South Carolina Mathematics Curriculum Standards.

Both the **Mid-Year PACT Test Prep** and the **Final PACT Test Prep** focus on the South Carolina Mathematics Curriculum Standards so that students can review critical state content as they hone their test-taking skills in each of the PACT formats: multiple-choice, open-response, and open-ended.

PACT Test Prep items also appear in South Carolina-themed lessons. These lessons provide additional coverage, practice, and review of South Carolina Curriculum Standards. The lesson numbers indicate when to use these pages. For example, you can use Lesson 1.4A anytime after Lesson 1.4 in your *Harcourt Math* program.

Problem Solving On Location offers real-life applications set in South Carolina locales and serve as rewarding, standard-based options for wrapping up a unit in *Harcourt Math.*

Table of Contents
Pupil Pages

Teacher's Pages

Choose the best answer.

1. Mrs. Bennett distributed 326 pencils to the students each month at the library. Which is the most reasonable estimate for the number of pencils used in one year?

   A 4,000   **C** 9,000
   B 6,000   **D** 30,000

2. The pastry chef used 56 pounds of flour at the restaurant each week. If she used the same amount every night, write an equation to find n, or the number of pounds of flour she used per night.

   F $n \times 56 = 56$ **H** $n \times 7 = 56$
   G $56 \div 8 = n$ **J** $8 \times n = 56$

3. If you know that $35 \times 40 = 1,400$, how can you find 35×80?

   A Write a zero after the product.
   B Find half of 1,400.
   C Write the same product.
   D Double 1,400.

4. What is 3,077,590 rounded to the nearest hundred thousand?

   F 3,000,000
   G 3,100,000
   H 3,200,000
   J 4,000,000

5. Rachel bought a pair of skates for $22. Her sister bought a pair for $47. Which estimated difference is closer to the exact answer?

   A $2 **B** $15 **C** $30 **D** $45

6. In the next six days, Lexi wants to run a total of 18 laps. She wants to run the same number of laps each day. Which equation can she use to solve the problem?

   F $18 + 6 = n$ **H** $18 - 6 = n$
   G $6 \times n = 18$ **J** $6 \times 18 = n$

For 7–8, use the bar graph.

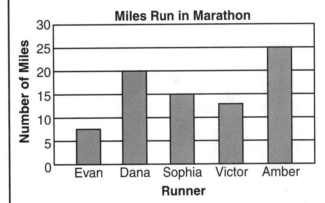

7. About how many more miles did Sophia run than Victor?

   A 10 mi   **C** 4 mi
   B 6 mi   **D** 2 mi

8. Which is the best estimate of the total number of miles the marathoners ran?

   F 80 mi   **H** 65 mi
   G 70 mi   **J** 50 mi

Go On

For 9–10, use the table.

FOURTH GRADE FIELD TRIP SCHEDULE	
Bus Departs	7:30 A.M.
Enter Park	9:00 A.M.
Lunch	12:30 P.M.
Leave Park	4:45 P.M.
Arrive at School	6:15 P.M.

9. The fourth-grade class went to a water park for a field trip. This is a schedule for the trip. The class had to leave the park 1 hour 15 minutes early due to a rainstorm. At what time did the class leave the park?

  A 2:15 P.M.   **C** 3:30 P.M.
  B 2:30 P.M.   **D** 3:45 P.M.

10. The bus was early arriving back at school. Use the clock below to read the time the students arrived back at school.

  F 25 minutes after four
  G 25 minutes after six
  H 20 minutes after five
  J 25 minutes to five

11. What is 7,905,410 written in expanded form?

  A 700,000 + 900,000 + 5,000 + 400 + 10
  B 7,000,000 + 9,000 + 5,000 + 400 + 10
  C 7,000,000 + 900,000 + 500 + 400 + 10
  D 7,000,000 + 900,000 + 5,000 + 400 + 10

12. Sarah is participating in a dance competition. She will dance in five different routines for a total of 1 hour 30 minutes. If each routine takes the same amount of time, how much time will each take?

  F 8 min   **H** 12 min
  G 10 min   **J** 18 min

13. Rebecca wants to calculate how much it would cost to buy 17 tickets to a soccer game at $5 per ticket. Which expression shows a way to calculate 17 × 5 mentally?

  A (7 × $5) + 10
  B (10 × $5) + (7 × $5)
  C 10 × 7 × $5
  D 10 × $5 + 7

© Harcourt

Go On ▶

For 14–15, use the line graph.

Alison's Bank Account

14. Which best describes the change in Alison's bank account between weeks 4 and 5?

F Alison put more money into her bank account.

G Alison took money out of her bank account.

H Alison did not go to the bank.

J Alison gave money to the bank teller.

15. What do you notice about the change in the number of dollars saved between weeks 1 and 5?

A The number of dollars saved decreases, then increases.

B The number of dollars saved increases.

C The number of dollars saved increases, then decreases.

D The number of dollars saved decreases.

16. How many periods are in a 7 digit number?

F 4   **G** 3   **H** 2   **J** 1

17. The book fair committee purchased 75 pocket folders at $1.25 each. Choose the equation to find the total cost.

A $c \times 75 = \$1.25$

B $c = 75 \times \$1.25$

C $c = 75 - \$1.25$

D $c \times \$1.25 = 75$

For 18, use the table.

FAVORITE INSTRUMENTS	
clarinet	8
flute	5
trumpet	5
oboe	3
percussion	10

18. Justin took a vote in his school's band of students' favorite instruments. Which instrument represents the mode?

F percussion   **H** clarinet

G flute   **J** trumpet

19. What is the value of the digit 4 in the number 5,435,650,000?

A 4,000   **C** 40,000

B 400,000,000   **D** 4,000,000,000

20. The heartbeat of an average dog weighing 30 pounds is about 100 beats per minute. At that rate, how many times would a dog's heart beat in 20 minutes?

F 200,000   **H** 2,000

G 20,000   **J** 200

Go On ▶

21. The Company Theater holds 462 people. Over five days, there will be seven performances of an amateur play. If each play is sold out, how many tickets have been sold? Tell which place-value positions need to be regrouped. [1 point]

_____

_____

22. What method would you use—paper and pencil or mental math—to compute 830×50? Explain. [1 point]

_____

_____

23. Ross records one science test score every week for 8 weeks. What kind of graph would be best to display this data? Explain. [1 point]

_____

_____

24. The Emperor penguin is the largest penguin that weighs about 80 pounds. How many pounds does a pack of 18 penguins weigh? Describe how you can make a model using grid paper to find the product 80×18. Tell what the product is. [1 point]

_____

_____

_____

25. When marine turtles lay eggs, they usually lay them in batches or clutches of six. Each clutch contains about 100 eggs. If there are 4 female marine turtles, about how many eggs will they lay? [1 point]

_____

Go On ➡

● **For 26, use the table.**

INPUT	p	3	4	5	6	7
OUTPUT	r	$18	$24	■	$36	$42

26. At Wild Run State Park, it costs $18 to rent a canoe for 3 hours. Brian and Alan want to go canoeing for 5 hours. How much will they have to pay? Find a rule for the table and name the missing number. Then write the equation.
[2 points]

_____

_____

For 27, use the bar graphs.

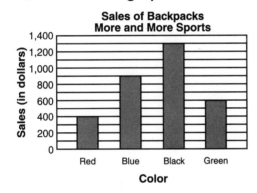

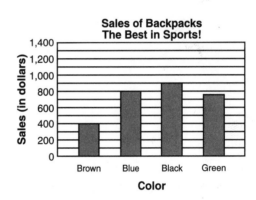

27. Ryan's job at Action Sports Manufacturing Company is to analyze the sales of backpacks at two leading sports stores. The graphs show the data he collected. Compare the shapes of the graphs. What conclusions can you draw from the sales data? Describe how Ryan can use the results.
[3 points]

_____

_____

_____

_____

_____

_____

Go On ▶

28. Mr. Sheridan's class took a survey of fourth grade students to find their favorite types of movies. The table below has the survey's results. Make a bar graph that shows the results of the survey. Be sure to

   • give your graph a title
   • use a scale
   • tell what each bar represents
   • write two statements that compare the data on types of movies. [3 points]

TYPE OF MOVIE	NUMBER OF STUDENTS
Cartoon	14
Comedy	21
Science Fiction	9
Adventure	16

_____

_____

_____

© Harcourt

Stop ▶

Choose the best answer.

1. Betty left for school at 8:23 A.M. She arrived at 9:00 A.M. How many minutes did it take her to get to school?

   A 23 min   **C** 47 min

   B 37 min   **D** 53 min

2. There are 3 quarts of punch. How many 1-cup servings will 3 quarts make?

1 quart = 2 pints
1 pint = 2 cups

   F 3 cups   **H** 6 cups

   G 4 cups   **J** 12 cups

For 3, use the solid figures.

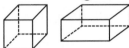

3. Which of the following statements is true?

   A A cube has more faces than a rectangular prism.

   B A cube has the same number of faces as a rectangular prism.

   C A cube has fewer faces than a rectangular prism.

   D The number of faces depends on the size of the cube or rectangular prism.

4. There are 94 students separated into equal groups and 4 students left over. Each group has more than 15, but fewer than 20 members. How many groups are there?

   F 3   **H** 5

   G 4   **J** 6

5. What are the factors of 49?

   A 1, 7, 49

   B 1, 7, 9, 49

   C 1, 3, 7, 9, 49

   D 2, 49, 98

For 6, use the spinner.

6. The spinner has four equally likely outcomes. What is the probability of the pointer stopping on 4?

   F $\frac{1}{8}$ **H** $\frac{1}{4}$

   G $\frac{1}{6}$ **J** $\frac{1}{2}$

For 7, use the drawing.

door

7. Jan was facing the door of her classroom. She turned to her right until her back was facing the door. How many degrees did she turn?

   A 45°   **C** 180°

   B 90°   **D** 270°

8. In science class, four of the students have plants with heights 4.25 cm, 5.36 cm, 4.52 cm, and 5.63 cm. Which statement is true?

   F 4.25 cm is greater than 5.36 cm

   G 4.25 cm is less than 5.36 cm

   H 5.63 cm is less than 5.36 cm

   J 4.25 cm is greater than 4.52 cm

Go On

For 9, use the model.

9. What decimal and fraction are shown by the model?

   A 25; $\frac{25}{1}$ **C** 0.25; $\frac{25}{100}$

   B 2.5; $2\frac{5}{10}$ **D** 0.025; $\frac{25}{100}$

10. Edward lives 3.9 km from the library. Sandy lives 500 m more than twice that distance from the library. How far from the library does Sandy live?

   F 8.3 km   **H** 7.3 km

   G 7.8 km   **J** 4.5 km

For 11, use the figure.

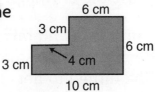

11. What is the area?

   A 36 sq cm   **C** 45 sq cm
   B 40 sq cm   **D** 48 sq cm

12. Which of the following shows a flip?

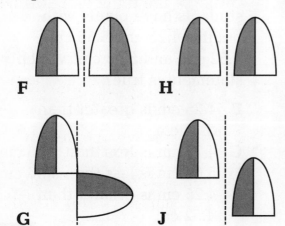

   F

   H

   G

   J

13. Matt and Jennie are playing a game called *Name the Pattern Rule*. Matt names these numbers: 2.045, 6.135, 18.405, 55.215. Jennie names a rule for the pattern. Which rule does she name?

   A Add 4.09.   **C** Multiply by 2.
   B Add 12.27.   **D** Multiply by 3.

For 14, use the drawing.

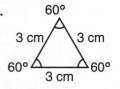

14. What terms describe this triangle by sides and by the measure of its angles?

   F scalene, obtuse
   G equilateral, acute
   H isosceles, right
   J equilateral, obtuse

For 15, use the figure.

15. What is the best estimate of the area of the polygon drawn on the grid?

   A 28 sq units   **C** 40 sq units
   B 32 sq units   **D** 62 sq units

16. Which is a solid figure with 3 rectangular faces and 2 triangular faces?

   F cube   **H** rectangular prism

   G cylinder   **J** triangular prism

Go On

For 17, use the model.

| $\frac{1}{5}$ | $\frac{1}{5}$ | $\frac{1}{5}$ | $\frac{1}{5}$ | $\frac{1}{5}$ |
| $\frac{1}{3}$ | | $\frac{1}{3}$ | | $\frac{1}{3}$ |

17. Compare the fractions. Which is true?

A $\frac{4}{5} < \frac{2}{3}$ **C** $\frac{2}{3} > \frac{4}{5}$

B $\frac{4}{5} > \frac{2}{3}$ **D** $\frac{2}{3} = \frac{4}{5}$

For 18, answer the question and tell how to interpret the remainder.

18. A 61-inch piece of ribbon needs to be cut into 8-inch lengths. How many 8-inch lengths will there be?

F 8; increase the quotient by 1

G 7; drop the remainder

H 7; use the remainder as the answer

J 5; use the remainder as the answer

19. Charles read the temperature at 7 P.M. at 25°F. During the night, the low temperature was ⁻5°F. By how many degrees did the temperature drop?

A 40° **B** 35° **C** 30° **D** 20°

20. Brittany's favorite TV program comes on at 4:00 P.M. John's favorite program is on $\frac{1}{2}$ hour later and lasts $1\frac{1}{2}$ hours. At what time does John's favorite program end?

F 4:30 P.M.   **H** 5:30 P.M.

G 5:00 P.M.   **J** 6:00 P.M.

21. Sheila has a number cube that is labeled 2, 2, 4, 4, 6, and 6. What is the probability that Shelia will toss a 2 or a 4?

A $\frac{5}{6}$ **B** $\frac{4}{6}$ **C** $\frac{3}{6}$ **D** $\frac{2}{6}$

For 22, use the model.

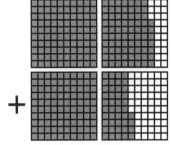

22. What is the sum of 1.77 + 1.44?

F 3.21   **H** 2.21

G 3.11   **J** 2.11

For 23–24, use the grid.

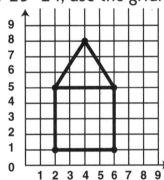

23. What is the ordered pair for the corner of the triangle that is **not** a corner of the square?

A (2,1) **B** (2,5) **C** (4,8) **D** (6,5)

24. Start at (4,9). Which describes how to move to (6,1)?

F 4 units down, 2 units left

G 2 units right, 8 units down

H 8 units down, 2 units left

J 4 units down, 2 units right

Go On ▶

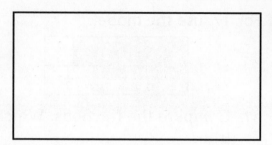
25. Eric has 15 tomato plants. He wants the plants to be 3 feet apart and also 3 feet from the garden fence. Draw a diagram of his garden. What is a reasonable area for Eric's garden? Explain. [2 points]

_____

_____

_____

For 26, use the drawing at the right.

26. Janie drew this figure. How can she draw this shape so that it shows a slide? How can she draw the original shape so that it shows a flip? Explain. [1 point]

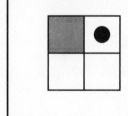

Slide

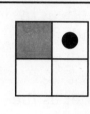

Flip

_____

_____

27. The Ruiz family visited the Columbia Museum of Art in Columbia, South Carolina. They saw points, lines, line segments, rays, and angles in many of the paintings. Draw a picture to show an example of each term. Label your drawings with the terms that describe them. [1 point]

Go On

● **For 28, use the figure.**

28. Divide the figure into 4 equal parts. Find the area of the whole figure. Then shade $\frac{1}{2}$ of the figure. Explain how you know that the parts you shaded are $\frac{1}{2}$ of the whole figure. [3 points]

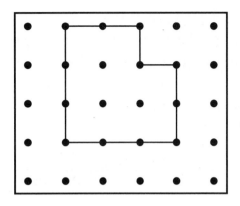

_____

_____

_____

29. Explain what each tool might be used to measure. [2 points]

●

_____

_____

_____

Felicia is making pancakes. Which tool should she use to measure the milk she needs in this recipe?

Pancakes

1 c pancake mix

1 egg

$\frac{3}{4}$ c milk

1 tsp vanilla

● _____

Name _____

30. Suppose Mary spins the pointer on each spinner. Make a tree diagram to list all the possible outcomes for the event. [3 points]

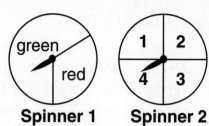

Spinner 1   **Spinner 2**

Tree diagram

Spinner 1	Spinner 2	Outcomes

Which outcomes are equally likely? Which outcomes are more likely? Explain your answer.

_____

_____

_____

_____

_____

Stop

Place Value Through Billions

 Learn

In 1998, the value of products produced and services performed in South Carolina was $100,350,000,000. If you counted one dollar every second, it would take almost 3,180 years to reach $100,350,000,000!

A place-value chart can help you read and write greater numbers. The period to the left of *millions* is **billions**.

VOCABULARY
billions

PERIOD

BILLIONS			MILLIONS			THOUSANDS			ONES		
Hundreds	Tens	Ones	Hundreds	Tens	Ones	Hundreds	Tens	Ones	Hundreds	Tens	Ones
1	0	0,	3	5	0,	0	0	0,	0	0	0

Standard Form: 100,350,000,000

Word Form: one hundred billion, three hundred fifty million

Expanded Form: 100,000,000,000 + 300,000,000 + 50,000,000

EXAMPLES

Standard Form	Word Form	Expanded Form
A 6,033,200,000	six billion, thirty-three million, two hundred thousand	6,000,000,000 + 30,000,000 + 3,000,000 + 200,000
B 50,150,000,905	fifty billion, one hundred fifty million, nine hundred five	50,000,000,000 + 100,000,000 + 50,000,000 + 900 + 5

Check

1. Tell how many digits are in the number ten billion. _____

Write the value of the digit 8 in each number.

2. 38,320,400,090   **3.** 846,390,000,290   **4.** 9,822,453,777

_____   _____   _____

Write the value of the underlined digit.

5. 5<u>3</u>4,888,245,454   6. <u>5</u>,800,000,050   7. 65,<u>8</u>75,500,000

_____   _____   _____

Write each number in standard form.

8. three hundred billion, four hundred thousand   _____

9. one billion, fifty million   _____

Write each number in word form.

10. 40,000,000,000   11. 6,500,000,000

_____   _____

12. Complete. Explain how place value and period names help you read and write numbers.

5,002,500,000 = 5 _____ + 2 _____ + 5 hundred thousand

_____

PACT Test Prep

For 13–14, use the table.

13. What is the value of the digit 1 in the amount for transportation?

   A $1,000,000,000   C $100,000
   B $1,000,000   D $1,000

14. If Health receives one hundred million dollars more in next year's budget, what will its budget be? Write this amount in standard form.

_____

SOUTH CAROLINA STATE BUDGET 2000–2001 (rounded to the nearest thousand)	
Department	**Amount**
Health	$ 5,194,564,000
K-12 Education	$ 2,805,620,000
Higher Education	$ 2,263,941,000
Other	$ 1,847,265,000
Transportation	$ 1,027,802,000
Public Safety	$ 763,907,000
Total	**$13,889,210,000**

© Harcourt

Name _____

PROBLEM SOLVING ON LOCATION

in South Carolina's Cities

In 1786, the capital of South Carolina was moved from the coastal city of Charleston to a more central location. The capital was named Columbia. The population of Columbia in 1786 was about 1,000. The table shows the 2000 population of Columbia and nine other South Carolina cities.

For 1–6, use the table.

1. Which South Carolina city has the greatest population?

   _____

2. What is the population of Charleston rounded to the nearest ten thousand?

   _____

3. Which city has the greater population—Mount Pleasant or Sumter? Explain.

   _____

4. Which cities have a population greater than 50,000 people?

TEN LARGEST CITIES IN SOUTH CAROLINA	
City	**Population (2000)**
Charleston	96,650
Columbia	116,278
Florence	30,248
Greenville	56,002
Hilton Head Island	33,862
Mount Pleasant	47,609
North Charleston	79,641
Rock Hill	49,765
Spartanburg	39,673
Sumter	39,643

_____

5. Estimate the number of people who live in the three largest cities. Name the cities.

_____

6. About how many more people live in Columbia in 2000 than in 1786?

_____

South Carolina has grown rapidly in the last 50 years, and currently ranks 26th in population among the 50 states. The table below shows South Carolina's increase in population since 1950.

For 7–15, use the table.

7. Round the state population for 2000 to the nearest million.

_____

SOUTH CAROLINA STATE POPULATION	
Year	Population
1950	2,117,027
1960	2,382,594
1970	2,590,516
1980	3,121,820
1990	3,486,703
2000	4,012,012

8. Describe how the size of the population changed between the years 1950 and 2000.

_____

9. If the population data for 1980 and 1990 are rounded to the nearest million, are the rounded numbers the same? Explain.

_____

10. Was there a greater population growth between 1980 and 1990 or between 1990 and 2000?

_____

11. What is the value of the digit 9 in the population for 1970?

_____

12. What number is three hundred fifty-six thousand, four hundred nine greater than the state population in 2000?

_____

13. How much greater is the 1980 population than the population for 1970? 1960?

_____

14. How can you write South Carolina's population for 2000 in word form?

_____

15. Write South Carolina's population for 1990 in expanded form.

_____

Compare Graphs

▶ **Learn**

Look at the data for Charleston, South Carolina and Abilene, Texas. What does the shape of these data show about the number of days of precipitation in each city?

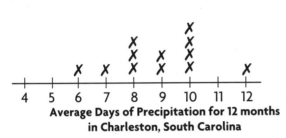

Average Days of Precipitation for 12 months in Charleston, South Carolina

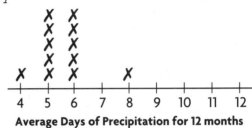

Average Days of Precipitation for 12 months in Abilene, Texas

Most of these data form a cluster at 8 to 10 days of precipitation.

Most of these data form a cluster at 5 to 6 days of precipitation.

So, the shape of the data in Charleston shows that most months have 8 to 10 days of precipitation. In Abilene, most months have 5 to 6 days of precipitation.

MATH IDEA Understanding the shape of the data will help you compare graphs and answer questions.

EXAMPLE

Compare the shapes of the graphs below. Which city has the greater decrease in days from March to April?

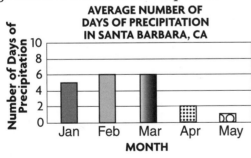

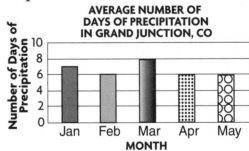

The bar for April is much shorter for Santa Barbara. So, Santa Barbara has the greater decrease.

▶ **Check**

1. **Explain** how the shape of data helps you compare two graphs.

_____

© Harcourt

For 2–3, use the line plots.

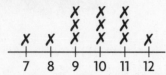

**Average Days of Precipitation for
12 months in Greenville-Spartanburg, SC**

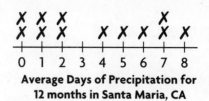

**Average Days of Precipitation for
12 months in Santa Maria, CA**

2. Which city has more months with days of precipitation?

_____

3. Which city has 2 months with seven days of precipitation?

_____

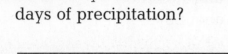
Practice and Problem Solving

For 4, use the bar graph.

4. Carol thinks that she will get 80 or above for her overall test grade. Explain how you know she is correct.

_____

_____

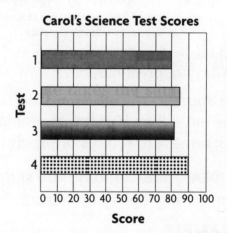

Carol's Science Test Scores

PACT Test Prep

For 5–6, use the line graph.

5. In which region is the average temperature for April 65°F?

_____

6. Compare the shapes of the data. Which statement is true for both regions?
   A The temperatures decrease.
   B The temperatures increase.
   C The temperatures stay the same.
   D The temperatures increase and decrease.

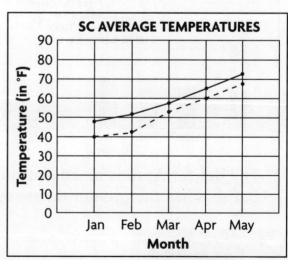

SC AVERAGE TEMPERATURES

Key: ——— Coastal Region
 - - - - Upstate Region

© Harcourt

Name _____

PROBLEM SOLVING ON LOCATION

in South Carolina's Blue Ridge Mountains

The Blue Ridge Mountains extend from northern Georgia to Pennsylvania and cover parts of six states. The name of the mountain range comes from the blue tone of the evergreen trees when they are seen from a distance.

At 3,560 feet, Sassafras Mountain is South Carolina's highest point.

For 1–5, use the table.

1. Which graph would better display the type of data in the table—a line graph or a bar graph? Why?

   _____

   _____

2. Order the mountain heights from highest to lowest.

   _____

   _____

3. Suppose a hiking trip on Sassafras Mountain is two days long. How many hours is this?

   _____

4. How many feet higher is Mount Mitchell than Mount Rogers?

PEAK HEIGHTS BLUE RIDGE MOUNTAINS	
Mountain	**Height (in feet)**
Sassafras	3,560
Mount Mitchell	6,684
Grandfather	5,964
Mount Rogers	5,729
Hawksbill	4,049
Brasstown Bald	4,784

_____

5. Which two mountains, when their heights are rounded to the nearest hundred, have a difference of 2,000 feet?

_____

UPCOUNTRY WATERFALLS

In places, the steep Blue Ridge Mountains are accessible only along narrow passes cut by cliff-walled rivers. Some of these rivers cascade hundreds of feet down mountain waterfalls. The bar graph shows the heights of some Upcountry waterfalls.

For 6–10, use the bar graph.

6. What is the range of the data?

   _____

7. What is the median waterfall height? Which waterfall is that tall?

   _____

8. What value could be considered an outlier? Explain.

   _____

   _____

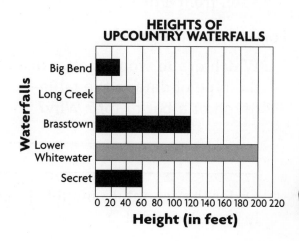

HEIGHTS OF UPCOUNTRY WATERFALLS

9. How would the lengths of the bars change if the interval were 10? If it were 25?

   _____

10. Which waterfall is twice as tall as Secret Falls?

    _____

11. Cristina's family took a 7-day camping and hiking trip to the Blue Ridge Mountains. Their trip began on April 28. On what day did the trip end?

    _____

12. During the trip, Cristina took a 2-hour hike to Secret Falls. If she left for the falls at 11:30 A.M., what time did she arrive there?

    _____

Identify the Missing Number

 Learn

The Greenville Pavilion Ice Rink is one of two public indoor ice rinks in South Carolina. Each week, 4 pairs of skate laces are replaced. Find the number of laces replaced after 5 weeks.

INPUT	a	1	2	3	4	5
OUTPUT	c	4	8	12	16	■

You can find a rule and then use the rule to find the missing number in the table.

Pattern: Each output is the input multiplied by 4.
Rule: Multiply by 4.
Input: 4   Output: $5 \times 4 = 20$

So, the number of laces replaced after 5 weeks is 20.

EXAMPLES

Find a rule. Name the missing number in the table or pattern.

A

INPUT	x	18	24	30	36
OUTPUT	y	3	4	5	■

Look for a pattern. Test the pattern.

Pattern: Each output is the input divided by 6.

Rule: Divide by 6.

Input: 36   Output: $36 \div 6 = 6$

So, the missing number is 6.

B   1, 2, 4, 8, 16, ■

- Look for a pattern. Possible rule: add 1 or multiply by 2.

- Test the rules for each pair of numbers. Multiply by 2 matches the pattern.

- Use the rule to find the missing number.
  $16 \times 2 = 32$.

So, the missing number in the pattern is 32.

1. **Explain** the steps you would use to find a missing number in a number pattern.

_____

Find a rule.

2. 3, 9, 27, 81

3.

INPUT	r	25	30	35	40
OUTPUT	s	5	6	7	8

_____   _____

▶ **Practice and Problem Solving**

Find a rule.

4.

INPUT	x	8	9	10	11
OUTPUT	y	32	36	40	44

5.

INPUT	w	24	32	40	48
OUTPUT	x	3	4	5	6

_____   _____

Find a rule. Name the missing number.

6. 400, 200, 100, ■

7. 45, 54, 63, 72, ■

_____   _____

PACT Test Prep

8. This famous pattern is known as Fibonacci's sequence: 1, 1, 2, 3, 5, 8, 13, 21…. Find the rule. Use the rule to find the next number.

_____

_____

9. Which equation describes the rule for the table in Problem 5?

A $x = w \div 4$ **C** $x = w \times 8$

B $w = x \times 3$ **D** $x = w \div 8$

10. The table at the right shows the cost to rent ice skates at a skating rink. What is the cost of skate rentals for five people?

INPUT	a	1	2	3	4	5
OUTPUT	r	$3	$6	$9	$12	■

F $10   **H** $15

G $13   **J** $18

© Harcourt

PROBLEM SOLVING ON LOCATION

in South Carolina's River Basins

South Carolina has over 11,000 miles of rivers and streams. Each of these, from the cold streams in the mountains to the blackwater rivers of the coast, lies within one of the eight river basins that cover the state. The river basins vary in size and shape, and are the areas of land that drain into a lake, river, or stream.

For 1–5, use the table.

1. Which river basins have an area of about 3 million acres?

   _____

   _____

2. Could the Broad River Basin's area be 2,545,832? Explain.

   _____

   _____

3. Write an expression with a variable that can be used to find the combined area of the Saluda River Basin and another basin. What does the variable represent?

SOUTH CAROLINA'S RIVER BASINS	
River Basin	**Area in Acres (to the nearest million)**
Broad	3,000,000
Catawba-Wateree	2,000,000
Edisto	2,000,000
Pee Dee	5,000,000
Salkehatchie	2,000,000
Saluda	2,000,000
Santee	2,000,000
Savannah	3,000,000

_____

4. About how many acres do the eight basins cover?

_____

5. Write an expression you could use to find the estimated number of acres in the Santee, Saluda, and Edisto River Basins. Use addition.

_____

The Edisto River Basin covers over 2 million acres of various types of land. The Edisto River, named by the Native Americans who lived there, is one of the longest free-flowing blackwater rivers in the United States. It flows through about 250 miles of South Carolina's coastal plain.

For 6–8, use the table.

6. Which two types of land cover the same amount of area?

   _____

7. What if the size of the barren land were twice as large? Write and evaluate the expression.

   _____

   _____

8. How many acres does the urban, agricultural, and scrubland areas include? Write an equation with a variable and solve.

EDISTO RIVER BASIN	
Type of Land	**Area in Acres (to the nearest thousand)**
Urban	36,000
Agricultural	454,000
Scrubland	218,000
Barren	9,000
Forest	980,000
Forested wetland	222,000
Wetlands	40,000
Water	40,000

9. One weekend, Michael and his family canoed 18 miles down the Edisto River. They canoed the same distance each of the two days. Write an expression that shows how far they canoed each day. Then find the value.

   _____

10. In a clean-up project on the Edisto River, each student from a group of 9 collected 3 bags of trash. Write an equation that shows the total number of bags of trash, t, collected by the group. Solve the equation.

   _____

Factor Patterns

▶ Learn

At Huntington Beach Park in March, a bird watcher counted 3 nests with 4 baby birds in each nest. In April, he counted 7 nests with 4 baby birds in each nest. How did the total number of baby birds change from March to April?

Huntington Beach Park is one of South Carolina's best areas for watching birds. Terns, gulls, petrels, loons, sparrows, and many other birds nest in this area.

March: $4 \times 3 = 12$

Think: $7 > 3$ and $28 > 12$

April: $4 \times 7 = 28$

So, the bird watcher counted more birds in April.

MATH IDEA When one factor increases, the product increases. When one factor decreases, the product decreases.

EXAMPLES

A $8 \times 5 = 40$
$5 \times 5 = 25$

So, the product decreased.

B $6 \times 4 = 24$
$6 \times 6 = 36$

So, the product increased.

When one factor is doubled or halved, the product is doubled or halved.

C $3 \times 4 = 12$
$3 \times 8 = 24$

So, the product is doubled.

D $3 \times 10 = 30$
$3 \times 5 = 15$

So, the product is halved.

▶ Check

1. **Describe** what happens to the product when one of the factors changes from 4 to 5 and the other factor stays the same. Give an example.

Write whether the product *increases* or *decreases*.

2. $3 \times 4 = 12$
   $3 \times 6 = \blacksquare$

   _____

3. $5 \times 9 = 45$
   $5 \times 7 = \blacksquare$

   _____

4. $12 \times 5 = 60$
   $6 \times 5 = \blacksquare$

   _____

Find the unknown product.

5. $2 \times 8 = 16$
   $2 \times 4 = $ _____

6. $6 \times 7 = 42$
   $6 \times 9 = $ _____

7. $4 \times 6 = 24$
   $8 \times 6 = $ _____

▶ **Practice and Problem Solving**

Write whether the product *increases* or *decreases*.

8. $8 \times 4 = 32$
   $8 \times 8 = \blacksquare$

   _____

9. $7 \times 3 = 21$
   $9 \times 3 = \blacksquare$

   _____

10. $10 \times 10 = 100$
    $10 \times 1,000 = \blacksquare$

    _____

Find the unknown product.

11. $5 \times 6 = 30$
    $5 \times 12 = $ _____

12. $7 \times 6 = 42$
    $7 \times 3 = $ _____

13. $3 \times 2 = 6$
    $3 \times 14 = $ _____

PACT Test Prep

For 14–15, use the table.

Sightings of Redheaded Woodpeckers	
Day	**Number**
Thursday	9
Friday	6
Saturday	12
Sunday	7

14. On which day did Molly spot 2 bird nests that each contained 6 woodpeckers?

    A Thursday   **C** Saturday
    B Friday   **D** Sunday

15. What operation is used to find the total number of woodpeckers seen from Thursday through Sunday?

    F addition   **H** multiplication
    G subtraction   **J** division

16. Which statement best describes the factor pattern in the equations, $4 \times 5 = 20$ and $4 \times 10 = 40$?

    A Both factors stayed the same, so the product stayed the same.
    B Both factors decreased, so the product decreased.
    C One factor stayed the same, the other factor doubled, so the product doubled.
    D Both factors doubled, so the product doubled.

© Harcourt

Closer Estimates

▶ **Learn**

The New Charleston Lighthouse at Sullivan's Island, South Carolina, is 163 feet tall. About how many feet will the lighthouse keeper walk in 14 days if he goes up and down the stairs once each day?

The New Charleston Lighthouse is the only lighthouse in the United States that has an elevator.

Estimate: $163 \times 2 \times 14 = 163 \times 28 \rightarrow 200 \times 30 = 6,000$

So, the lighthouse keeper will walk about 6,000 feet.

You can find a closer estimate by rounding 163 to 160.

EXACT PRODUCT   **ESTIMATE**

$$\begin{array}{r} 163 \\ \times\ 28 \\ \hline 4,564 \end{array}$$ $$\begin{array}{r} 160 \\ \times\ 30 \\ \hline 4,800 \end{array}$$ So, a closer estimate is 4,800 feet.

Sometimes when you round factors to numbers that are really close to the factors, you can describe an estimate as a *little more* than or a *little less* than the exact product.

EXAMPLES

A 122×31

EXACT: $122 \times 31 = 3,782$

ESTIMATE: $120 \times 30 = 3,600 \leftarrow$ *a little less than 3,782*

B 279×19

EXACT: $279 \times 19 = 5,301$

ESTIMATE: $280 \times 20 = 5,600 \leftarrow$ *a little more than 5,301*

Another way to estimate is to find two estimates that an exact product is *between*.

EXAMPLE 266×52

LOW ESTIMATE: $200 \times 50 = 10,000$
   Round both factors down.

EXACT: $266 \times 52 = 13,832$

HIGH ESTIMATE: $300 \times 60 = 18,000$
   Round both factors up.

So, the product 266×52 is *between* 10,000 and 18,000.

© Harcourt

1. Explain two ways to estimate 317×42.

_____

Find an estimate that is closer to the exact product.

2. $128 \times 43 = 5,504$
    Estimate: $100 \times 40 = 4,000$
    Closer Estimate: _____

3. $25 \times 77 = 1,925$
    Estimate: $30 \times 100 = 3,000$
    Closer Estimate: _____

► **Practice and Problem Solving**

Find an estimate that is closer to the exact product.

4. $238 \times 22 = 5,236$
    Estimate: $200 \times 20 = 4,000$
    Closer Estimate: _____

5. $151 \times 38 = 5,738$
    Estimate: $200 \times 40 = 8,000$
    Closer Estimate: _____

Give two estimates that the exact product is between.

6. $34 \times 38 = 1,292$
    Low Estimate: _____
    High Estimate: _____

7. $27 \times 155 = 4,185$
    Low Estimate: _____
    High Estimate: _____

PACT Test Prep

For 8–9, use the table.

Lighthouse	Location	Tower Height	Year Built
Leamington	Hilton Head	94 ft	1880
New Charleston	Sullivan's Island	163 ft	1962

8. The keeper of the New Charleston Lighthouse walked the tower 10 times. Which equation shows an estimate that is a little more than the exact answer?

A $200 \times 10 = 2,000$
B $190 \times 10 = 1,900$
C $170 \times 10 = 1,700$
D $160 \times 10 = 1,600$

9. Suppose the keeper of the Leamington Lighthouse walked the tower 18 times. What estimate is between the low and high estimate?

F 650 ft   **H** 850 ft
G 800 ft   **J** 1,800 ft

© Harcourt

Name _____

PROBLEM SOLVING ON LOCATION

Along South Carolina's Coastline

LIGHTHOUSES

A lighthouse is a tower that sends a beam of light out to sea. It guides ships to navigate safely along coastlines and waterways. Lighthouses have been built for at least 3,000 years, and they are still used today.

This is the Georgetown Lighthouse located on North Island.

For 1–4, use the table.

1. The keeper rode the elevator in the New Charleston Lighthouse 12 times a day for 15 days. Estimate the total number of elevator trips by rounding one factor up and one factor down. How does the estimate compare to the product? _____

South Carolina's Tallest Lighthouses		
Lighthouse	**Location**	**Height (in ft)**
Cape Romain	McClellanville	150
Georgetown	North Island	87
Hunting Island	Hunting Island State Park	136
Leamington	Hilton Head	94
New Charleston	Sullivan's Island	163
Old Charleston	Morris Island	165

2. Estimate the total number of feet climbed, if the keeper of the Georgetown climbed to the top once a day for 28 days. How would the estimate change if the keeper climbed the Georgetown for 18 days? Explain.

_____

3. Order the heights of the six lighthouses from tallest to shortest.

_____

4. The keepers of the Old Charleston and Cape Romain estimated the number of feet climbed during a 30-day period. If each height is rounded up, which estimate is closer to the exact answer? Explain.

_____

© Harcourt

HUNTING ISLAND STATE PARK

The only lighthouse in South Carolina that is open to the public is the Hunting Island Lighthouse. This lighthouse, built in 1875, replaced the original 1859 structure. Hunting Island State Park was established at the site of the lighthouse in the 1930s, and today is a popular camping and beach destination.

HUNTING ISLAND STATE PARK CAMPING FEES		
	April–October	**November–March**
Campsite	$22/night	about $18/night
	October–April	**May – September**
Cabin	$234 for 3 nights	about $468 for 1 week

For 5–9, use the table.

5. At the park, there are a total of 200 campsites. Find the total fee for one summer night if all sites are rented.

_____

6. During the busy season, the average fee for renting a cabin for one week is $468. What is the cost for four weeks?

_____

7. A large group of campers reserved 29 tent sites for one night in November. Estimate the total fee for the group. Tell how you rounded each factor.

_____

8. How much does it cost to rent one campsite for two weeks during December?

_____

9. A cabin can sleep 6 to 10 people, while a campsite can accommodate a 2-person tent. For three nights in October, would it cost less for a family of 6 to rent one cabin or three campsites? Explain.

_____

© Harcourt

Compare the Quotient and the Dividend

 Learn

Which has a greater quotient, $36 \div 3$ or $52 \div 3$?

Look for a pattern in these division problems.

$\begin{array}{r} 5 \\ 3\overline{)15} \end{array}$	$\begin{array}{r} 6 \\ 3\overline{)18} \end{array}$	$\begin{array}{r} 7 \\ 3\overline{)21} \end{array}$	$\begin{array}{r} 8 \\ 3\overline{)24} \end{array}$

As the dividend increases, the quotient increases.

So, since $52 > 36$, then $52 \div 3 > 36 \div 3$.

Compare. Look at the divisor.

$48 \div \mathbf{4} = 12$	The dividend, 48, is 4 times as great as the quotient, 12.
$72 \div \mathbf{4} = 18$	The dividend, 72, is 4 times as great as the quotient, 18.
$96 \div \mathbf{4} = 24$	The dividend, 96, is 4 times as great as the quotient, 24.

- In each problem, what does the divisor, 4, tell you about the size of the dividend compared to the quotient?

EXAMPLES

A $\begin{array}{r} 12 \\ 5\overline{)60} \end{array}$

The dividend is 5 times as great as the quotient.

B $\begin{array}{r} 13 \\ 6\overline{)78} \end{array}$

The dividend is 6 times as great as the quotient.

C $\begin{array}{r} 14 \\ 7\overline{)98} \end{array}$

The dividend is 7 times as great as the quotient.

- In $72 \div 8 = 9$, the dividend is how many times as great as the quotient?

 Check

1. **Explain** how you can compare the size of the quotient to the dividend in a division problem. _____

_____

© Harcourt

For 2–4, complete the table.

dividend ÷ divisor	2. 91 ÷ 7	3. 84 ÷ 6	4. 75 ÷ 5
quotient	_____	_____	_____
	The dividend is _____ times as great as the _____ .	The _____ is _____ times as great as the quotient.	The _____ is 5 times as great as the _____ .

▶ Practice and Problem Solving

For 5–7, complete the table.

5. 48 ÷ 3 = 16	The dividend, _____, is _____ times as great as the quotient, _____.
6. 68 ÷ 4 = 17	The dividend, _____, is _____ times as great as the quotient, _____.
7. 90 ÷ 5 = 18	The dividend, _____, is _____ times as great as the quotient, _____.

PACT Test Prep

For 8–9, use the table at the right.

8. Ninety-eight fourth graders are visiting the zoo. The teachers divide the students into equal-sized groups so they can spend time at each different exhibit. How many students will be in each group?

   A 10   **C** 14

   B 12   **D** 16

EXHIBITS AT RIVERBANKS ZOO IN COLUMBIA, SC
African Plains
Aquarium/Reptile Complex
The Birdhouse
Large Mammals
Small Mammals
Sea Lions
Riverbanks Farm

9. The students spend 45 minutes at each exhibit and have a 30-minute lunch break. How long are they at the zoo?

   F 5 hr   **H** 5 hr 30 min

   G 5 hr 15 min   **J** 5 hr 45 min

10. Some students put 32 photos into 4 albums. Each album has the same number of photos. If the students had 36 photos, would the number of photos in the 4 albums increase or decrease? Explain.

_____

_____

Change Units of Time

▶ Learn

Roshanda and her family are driving from Charleston to Clinton. A travel guide says that the trip takes about 3 hours. The computer calculates that the trip will last 180 minutes. Are these amounts of time the same?

Units of Time
60 **seconds (sec)** = 1 minute (min)
60 minutes (min) = 1 hour (hr)
24 hours (hr) = 1 day

One Way   Multiply to change and compare units of time.

3 hours = ■ minutes   To change a larger unit to a smaller unit, multiply.

number of hours	×	number of minutes in 1 hour	=	total minutes
3	×	60	=	180

Another Way   Divide to change and compare units of time.

180 min = ■ hr   To change a smaller unit to a larger unit, divide.

number of minutes in 3 hours	÷	number of minutes in 1 hour	=	number of hours
180	÷	60	=	3

So, 3 hours is the same amount of time as 180 minutes.

• Explain how to find the number of seconds in 2 hours.

▶ Check

1. **Tell** how you would find the number of minutes in 4 hours. _____

Complete.

2. 5 hr = ■ min

3. 240 sec = ■ min

4. 1 hr = ■ sec

_____

_____

_____

▶ **Practice and Problem Solving**

Complete.

5. 120 sec = ■ min

6. 360 min = ■ hr

7. 48 hr = ■ days

_____

_____

_____

8. 2 hr = ■ min

9. 6 days = ■ hr

10. 300 sec = ■ min

_____

_____

_____

11. Louis and Tim left for the soccer game at 2:20 P.M. They arrived back home at 5:35 P.M. How long were they gone?

12. Anthony drives 75 minutes to his uncle's home in Myrtle Beach. The return trip takes 63 minutes. How long is the round trip?

_____

_____

PACT Test Prep

For 13–14, use the line graph.

13. In which month is the normal temperature the greatest?

A Jun

C Aug

B Jul

D Sep

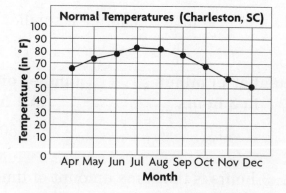

14. How much greater is the temperature in August than in April?

F 5°

H 31°

G 16°

J 42°

15. It takes Laura 25 minutes to wash her car and 12 minutes to dry it. What is the total amount of time in seconds?

A 62 sec

C 2,220 sec

B 2,000 sec

D 3,240 sec

© Harcourt

Common Multiples

VOCABULARY

common multiple

▶ **Learn**

South Carolina elects a governor every four years and a U.S. Senator every six years. In 2010 both positions will be up for election. In the next 60 years after 2010, how many times will the election for both positions occur at the same time?

You will need to find common multiples of 4 and 6. A **common multiple** is a multiple shared by two or more numbers.

Multiples of 4: 4, 8, **12**, 16, 20, **24**, 28, 32, **36**, 40, 44, **48**, 52, 56, **60**

Multiples of 6: 6, **12**, 18, **24**, 30, **36**, 42, **48**, 54, **60**

So, the election will occur at the same time 5 different times.

• What are the first three common multiples of 4 and 6?

ACTIVITY

• Look at the chart. Shade the numbers that are multiples of 3.

• Draw **X**s through the numbers that are multiples of 5.

1	2	3	4	5	6	7	8	9	10
11	12	13	14	15	16	17	18	19	20
21	22	23	24	25	26	27	28	29	30

• Look for the shaded squares that also have an X. Which numbers are common multiples of 3 and 5?

▶ **Check**

1. **Tell** how to find a common multiple of 2 and 5. _____

_____

Find three common multiples for each pair of numbers.

2. 3 and 8 _____   3. 4 and 12 _____   4. 6 and 10 _____

Practice and Problem Solving

Find three common multiples for each pair of numbers.

5. 3 and 7 ________

6. 8 and 10 ________

7. 5 and 7 ________

8. 3 and 6 ________

9. 1 and 5 ________

10. 9 and 12 ________

11. Cheryl found some common multiples of 8 and another number. The first three common multiples are 24, 48, and 72. What is the other number?

____

12. Two hundred eighty-two students are taking a field trip to the state capital. They are divided equally into 6 buses. How many students ride in each bus?

____

PACT Test Prep

For 13–16, use the table.

SOUTH CAROLINA ELECTED OFFICIALS	
Office	**Term**
Governor	4 years
State Senator	4 years
State House Rep.	2 years
U.S. Senate	6 years
U.S. House Rep.	2 years

13. If the election for governor and state house rep. occur in 2006, in what year will they both occur again?

A 2008 **C** 2012

B 2010 **D** 2016

14. If a South Carolinian is elected to 4 state house rep. and 2 state senate terms, how many years will she serve?

F 6 **H** 12

G 8 **J** 16

15. If the state senators and U.S. senators are elected in the same year, how many years later will both elections occur in the same year?

A 4 **C** 8

B 6 **D** 12

16. Which official will serve the longest period of time—a governor who serves 2 terms, a state senator who serves 3 terms, or a U.S. senator who serves 2 terms?

____

Name _____

PROBLEM SOLVING ON LOCATION

at South Carolina's Festivals

FESTIVAL OF ROSES

The South Carolina Festival of Roses takes place at the Edisto Memorial Gardens in Orangeburg. The 3-day festival includes exhibits, performances, music, and competitions. Some of the tournaments during this festival include softball, golf, basketball, horseshoes, and tennis.

FESTIVAL OF ROSES SCHEDULE			
	Friday	**Saturday**	**Sunday**
People Movers Shuttle	10:00 A.M.–5:00 P.M.	9:00 A.M.–6:00 P.M.	Noon–6:00 P.M.
Arts & Crafts Exhibit	Noon–5:00 P.M.	9:00 A.M.–6:00 P.M.	Noon–6:00 P.M.
Sonshine the Clown		11:00 A.M.–4:00 P.M.	2:00 P.M.–4:00 P.M.
Duck Race		9:00 A.M.–6:00 P.M.	Noon–6:00 P.M.

For 1–3, use the schedule.

1. How many hours is Sonshine the Clown performing in all? How many minutes is this?

   _____

2. Julie says that the People Movers Shuttle will run for a total of 540 minutes on Saturday. Is she correct? Explain.

   _____

3. In the duck race, plastic ducks float down a river to a finish line. If a duck race lasts about 12 minutes, how many duck races could take place in one hour? Write an equation and solve.

   _____

4. Riverside Drive was closed for the festival from 7 A.M. on Friday until 6:30 P.M. on Sunday. How long was Riverside Drive closed for the festival?

   _____

SPOLETO FESTIVAL USA

The Spoleto Festival USA takes place in Charleston, South Carolina. More than 100 performances of dance, music, and theater can be seen in locations throughout the city. Over 60,000 people attend this 17-day festival each year to see a variety of performances and art exhibits by artists from around the world.

5. The Spoleto Festival USA begins on May 25, 2001. What is the date of the last day of the festival?

_____

6. During the festival, opera productions can be seen in the Gaillard Municipal Auditorium. It is the largest indoor theater in Charleston and can seat 2,734 people. What is the greatest number of people that could attend 3 productions in the auditorium?

_____

7. Evening jazz concerts during Spoleto can be seen on the campus of the College of Charleston. This college was founded in 1770. How many years ago was the College of Charleston founded?

_____

8. At Spoleto, a dance production is scheduled to be in the Emmett Robinson Theater. This theater has 290 seats. A tour group traveling on 4 buses is going to the show. There are 43 people on each bus. How many theater seats will the group need? How many seats will be available for other people?

_____

9. An orchestra's rehearsal is at 10:35 A.M. and lasts for 1 hour 15 minutes. The orchestra then takes a break before the performance which begins at 1:00 P.M. How much time is there from the end of rehearsal to the start of the performance?

_____

10. **REASONING** If you know the number of people who attended each of 3 ballet performances, explain how you can find the mean.

_____

Name _____

Classify Triangles by Angles

 Learn

In this lesson you will learn how to classify triangles by the measures of their angles.

On a map, South Carolina's shape is similar to a triangle. Look at the triangle. What kind of triangle is outlined on South Carolina's borders?

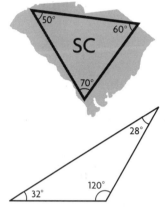

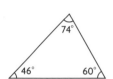

An **acute triangle** has 3 acute angles.

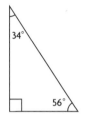

A **right triangle** has one right angle.

An **obtuse triangle** has one obtuse angle.

All three angles in the outlined triangle are acute angles. So, the triangle is an acute triangle.

MATH IDEA You can classify triangles by the lengths of their sides and by the measures of their angles.

ACTIVITY

Materials: centimeter ruler, protractor

• Measure the lengths of the sides of each triangle to the nearest centimeter. Record your measurements.

• Measure the size of the angles for each triangle. Record your measurements.

a.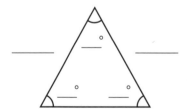

b.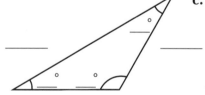

c.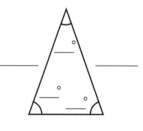

• Classify each triangle by the length of its sides and by the measures of its angles.

a. _____ b. _____ c. _____

1. **Explain** how you can classify triangles by the measures of their angles. _____

_____

Classify each triangle. Write *acute*, *right*, or *obtuse*.

2.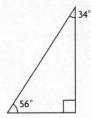

_____

3.

_____

4.

_____

► **Practice and Problem Solving**

Classify each triangle. Write *acute*, *right*, or *obtuse*.

5.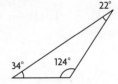

_____

6.

_____

7.

_____

Classify each triangle by the length of its sides (*equilateral, isosceles, scalene*) and the measures of its angles (*obtuse, right, acute*).

8.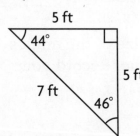

_____

9.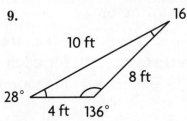

_____

10.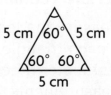

_____

PACT Test Prep

11. Draw a square and one of its diagonals. Classify the triangles.

    A right isosceles   **C** acute obtuse

    B right obtuse   **D** acute isosceles

12. Can a triangle be both scalene and right? Explain.

_____

_____

Name _____

PROBLEM SOLVING ON LOCATION

at the Columbia Museum of Art

MODERN ARCHITECTURE

The Columbia Museum of Art in Columbia, South Carolina displays paintings and sculptures of the last 600 years from around the world. Its new facility, which opened in 1998, has over 20,000 square feet of gallery space, and was built using many different geometric shapes.

For 1–7, use the photograph of the entrance hall at the Columbia Museum of Art.

1. Describe the relationship between $\overline{DE}$ and $\overline{FG}$.

_____

2. Classify triangle *ABC*. Write *isosceles*, *scalene*, or *equilateral*.

_____

3. Trace $\angle CAB$. Then use a protractor to measure the angle.

_____

4. What kind of angle is $\angle ABC$? Write *right*, *acute*, or *obtuse*.

_____

5. Tell whether triangle *ABC* has *rotational symmetry*, *line symmetry*, or *both*.

_____

6. Identify the shapes of Figures 1 and 2. Tell whether they are *congruent*, *similar*, or *neither*.

_____

7. Classify the figure *BCJH* in as many ways as possible. Write *quadrilateral*, *parallelogram*, *rhombus*, *rectangle*, *square*, or *trapezoid*.

_____

© Harcourt

CLASSIC ART

One of the new items at the Columbia Museum of Art is a French porcelain saucer made in 1790. It has a blue and orange enameled surface which displays a variety of Roman classical figures.

For 8–11, use the photograph of the saucer.

8. Are the different circles on the saucer *congruent*, *similar*, or *neither*? Explain.

   _____

   _____

9. Does the design in the center of the saucer have *line symmetry*, *rotational symmetry*, or *both*?

   _____

10. Estimate the circumference of the saucer if its radius is 3 inches.

    _____

11. The circumference of the inner circle on the saucer is about 6 inches. What is the diameter of the inner circle?

    _____

12. Many of the paintings on display at the museum are on rectangular canvases. Identify the kind of angle that makes up each of the four corners of a rectangular painting.

    _____

13. A painting at the museum is displayed in a plain rectangular frame. Tell what you know about the opposite sides of the frame.

    _____

14. A painter uses a canvas that has two pairs of parallel sides, with all four sides the same length. What figures can the canvas be?

    _____

Fractions on a Number Line

Learn

Tara and Maya are reading the same book. Tara has read $\frac{5}{8}$ of the pages and Maya has read $\frac{7}{8}$. Who has read half of the book? Who has read almost all of it?

You can use a number line to mark points to represent $\frac{5}{8}$ and $\frac{7}{8}$. Then, relate the size of these fractions to the benchmarks 0, $\frac{1}{2}$, and 1.

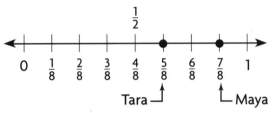

On the number line, 0 represents the beginning of the book, and 1 represents the end of the book.

Tara has read about half of the book because $\frac{5}{8}$ is close to $\frac{1}{2}$. The fraction $\frac{7}{8}$ is close to 1, so Maya has almost finished the book.

ACTIVITY   **Materials:** ruler

• Copy the number line below.

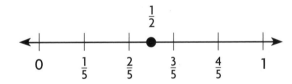

• Locate and graph points for $\frac{1}{5}$, $\frac{2}{5}$, $\frac{3}{5}$, and $\frac{4}{5}$.

• Tell whether each fraction is closest to 0, $\frac{1}{2}$, or 1.

MATH IDEA You can use a number line and benchmark fractions to understand and compare the size of fractions.

Check

1. **Tell** how you would decide if the fraction $\frac{5}{8}$ is closest to 0, $\frac{1}{2}$, or 1. _____

_____

For 2–3, use the number lines. Locate and graph points for $\frac{1}{6}$, $\frac{2}{6}$, $\frac{3}{6}$, $\frac{4}{6}$, $\frac{5}{6}$, and $\frac{1}{4}$, $\frac{2}{4}$, $\frac{3}{4}$.

2.

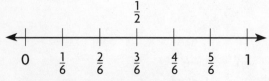

3.

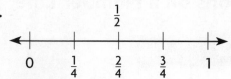

For 4–5, use the number lines. Locate and graph points for $\frac{1}{8}$, $\frac{2}{8}$, $\frac{3}{8}$, $\frac{4}{8}$, $\frac{5}{8}$, $\frac{6}{8}$, $\frac{7}{8}$, and $\frac{1}{7}$, $\frac{2}{7}$, $\frac{3}{7}$, $\frac{4}{7}$, $\frac{5}{7}$, $\frac{6}{7}$.

4.

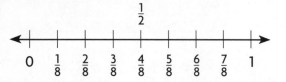

5.

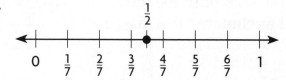

Write whether the fraction is closest to 0, $\frac{1}{2}$, or 1.

6. $\frac{4}{8}$ _____   7. $\frac{1}{8}$ _____   8. $\frac{5}{8}$ _____   9. $\frac{2}{7}$ _____   10. $\frac{6}{7}$ _____   11. $\frac{1}{7}$ _____

For 12–14, use the number line to identify the point for each.

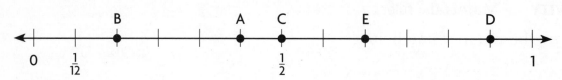

12. closest to 1 _____   13. shows $\frac{5}{12}$ _____   14. shows $\frac{2}{12}$ _____

PACT Test Prep

15. Brittany has traveled $\frac{9}{10}$ of the distance from Charleston to Columbia. Use benchmark fractions to choose the best description of the distance traveled.

   A She has just started.
   B She is halfway there.
   C She is almost there.
   D She has arrived.

16. Use a ruler to draw and label a number line divided into tenths. Include the benchmarks 0, $\frac{1}{2}$, and 1. Graph the points between 0 and 1.

_____

Compare Decimals

 Learn

One of the instruments used to
predict weather is a barometer.
A barometer measures air pressure.
The table shows the air pressures
for Columbia, South Carolina and
Greenville, South Carolina on
May 7, 2001. Which city had the
greater air pressure?

AIR PRESSURE in millimeters of mercury (May 7, 2001)	
City	**Pressure**
Columbia, SC	30.33
Greenville, SC	30.38

EXAMPLE

Use a place-value chart to compare the decimals.

Think: Line up the decimal points. Compare the digits, beginning with the greatest place value.

Tens	Ones	.	Tenths	Hundredths
3	0	.	3	3
3	0	.	3	8

$3 = 3$ $0 = 0$ $3 = 3$ $8 > 3$

Since $8 > 3$, 30.38 is greater than 30.33.

So, Greenville, South Carolina had the greater air pressure.

MORE EXAMPLES

Compare. Use *greater than*, *less than*, or *equal to*.

A 2.45 is __?__ 2.36.	**B** 0.34 is __?__ 0.38.	**C** 7.80 is __?__ 7.89.
2.45 is *greater than* 2.36.	0.34 is *less than* 0.38.	7.80 is *less than* 7.89.
D 5.18 is __?__ 5.17.	**E** 3.02 is __?__ 3.02.	**F** 0.02 is __?__ 0.06.
5.18 is *greater than* 5.17.	3.02 is *equal to* 3.02.	0.02 is *less than* 0.06.

© Harcourt

1. **Tell** how to compare the decimals 3.45 and 3.55.

_____

_____

Compare. Write *greater than, less than,* or *equal to.*

2. 5.7 is _____ 5.8

3. 12.4 is _____ 1.24

4. 0.05 is _____ 0.15

5. 27.09 is _____ 27.09

► **Practice and Problem Solving**

Compare. Write *greater than, less than,* or *equal to.*

6. 1.4 is _____ 2.4

7. 0.2 is _____ 0.3

8. 1.24 is _____ 1.14

9. 1.4 is _____ 1.04

10. 3.36 is _____ 3.38

PACT Test Prep

11. What are the missing decimals in the following pattern?

1.76, 2.035, 2.310, ■, ■, 3.135

A 2.035, 1.76
B 2.585, 2.860
C 2.585, 3.335
D 5.06, 7.81

12. Julie pays $186 each month for dance classes. She takes 6 different kinds of dance. To the nearest dollar, is the cost of each dance class *greater than, less than,* or *equal to* $30?

F greater than
G equal to
H less than

13. Glenn's bag weighs 24.75 pounds. Erin's bag weighs 24.25 pounds. Write *is greater than, is less than,* or *is equal to.*

The weight of Glenn's bag

_____

the weight of Erin's bag.

14. Compare the decimals. Which is true?

A 0.12 < 0.02   **C** 0.73 > 0.78
B 0.45 < 0.54   **D** 0.89 > 0.99

© Harcourt

Name _____

on South Carolina's Farms

Farmers in South Carolina grow dozens of different crops—some of which are shipped all over the country. One important crop is peaches. The table below shows the approximate fraction of South Carolina's peach trees that are located in each of three general regions.

South Carolina's Peach Trees	
Region	Fraction of Total Trees
Upper State	$\frac{1}{3}$
Ridge	$\frac{1}{2}$
Coastal Plains	$\frac{1}{6}$

For 1–6, use the table.

1. Order the regions from greatest number of trees to least.

   _____

2. What fraction of the trees are located in the Ridge and Coastal Plains regions together? _____

3. What fraction of the trees are located in the Upper State and Coastal Plains regions together? _____

4. Explain what picture you could draw to represent the fraction of trees in the Coastal Plains region.

   _____

5. How much greater is the fraction of trees in the Upper State region than in the Coastal Plains region? _____

6. Write two equivalent fractions for the fraction of trees in the Upper State region.

   _____

Peanuts are another important crop in South Carolina. They are grown in nearly every county in the state. This table shows how many pounds of peanuts were produced in each district in South Carolina in 1999.

1999 South Carolina Peanut Production	
District	Peanut Production (in millions of pounds)
Eastern	1.23
West Central	2.73
Central	11.34
Southern	9.97

For 7–12, use the table.

7. Which district had the greatest peanut production?

_____

8. Did the Eastern and Southern districts combined, produce a greater amount of peanuts than the Central district? Explain.

_____

9. How many millions of pounds more did the district with the greatest production produce than the district with the least production?

_____

10. Estimate the total production, in millions of pounds, for all four districts.

_____

11. Order the four districts from least to greatest production.

_____

12. What was the production, in millions of pounds, of the Southern district, rounded to the nearest tenth?

_____

Divide Shapes

 Learn

You can use line segments to divide figures and form new
figures. Look at the figures below.

A vertical line segment
divides the square into
two rectangles.

A horizontal line
segment divides the
rectangle into two
rectangles.

A diagonal line segment
divides the square into
two triangles.

When a figure is divided into congruent parts, each part can
be written as a fraction of the original figure. How does the
area of one part compare to the area of the original figure?

EXAMPLES

Find the area of the original figure and the area of the shaded
part of the divided figure. Then compare the areas. What
conclusion can you make?

A

3 in.
3 in.

$A = l \times w$
$A = 3 \times 3$
$A = 9$ sq in.

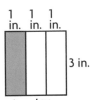
1 in. 1 in. 1 in.
3 in.

$A = l \times w$
$A = 1 \times 3$
$A = 3$ sq in.

Write a fraction in simplest form to
compare the areas:

$$\frac{3}{9} = \frac{3 \div 3}{9 \div 3} = \frac{1}{3}$$

Conclusion: Since the figure is
divided into 3 congruent parts,
the area of one part is $\frac{1}{3}$ the area
of the whole.

B
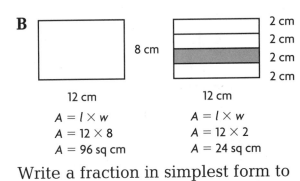
8 cm
12 cm

2 cm
2 cm
2 cm
2 cm
12 cm

$A = l \times w$
$A = 12 \times 8$
$A = 96$ sq cm

$A = l \times w$
$A = 12 \times 2$
$A = 24$ sq cm

Write a fraction in simplest form to
compare the areas:

$$\frac{24}{96} = \frac{24 \div 24}{96 \div 24} = \frac{1}{4}$$

Conclusion: Since the figure is
divided into 4 congruent parts,
the area of one part is $\frac{1}{4}$ the area
of the whole.

1. **Show** two ways to divide the rectangle into 4 congruent rectangles. Label the lengths and widths of your divided rectangles.

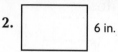

Find the area of the whole figure and the area of the shaded part. Write a fraction that compares the area of the shaded part to the whole.

2.   6 in.   ⬜ 2 in. / 2 in. / 2 in.   3.   10 cm   ⬜ 5 cm / 5 cm
   9 in.   9 in.   10 cm   5 cm 5 cm

A = _____   A = _____   A = _____   A = _____

Fraction: = _____   Fraction: = _____

► **Practice and Problem Solving**

Find the area of the whole figure and the area of the shaded part. Write a fraction that compares the area of the shaded part to the whole.

4.   8 in. / 4 in.   4 in. 4 in. / 2 in. / 2 in.   5.   8 m / 8 m   2 2 2 2 m m m m / 2 m / 2 m / 2 m / 2 m

A = _____   A = _____   A = _____   A = _____

Fraction: = _____   Fraction: = _____

PACT Test Prep

6. Draw a figure and divide it into congruent parts. Shade one of the parts. What conclusion can be made about the area of the shaded part compared to the area of the whole figure?

_____

7. A square with 4-inch sides is divided into two congruent rectangles. Find the area of one of the rectangles.

   A 8 sq in.   **B** 12 sq in.   **C** 16 sq in.   **D** 20 sq in.

Hands On • Patterns and Solid Figures

EXPLORE

An architect uses two-dimensional patterns to build models. The architect cuts out this pattern for a new project. What solid figure did the architect make?

ACTIVITY 1

Materials: pattern, scissors, tape

STEP 1	STEP 2
Cut out the pattern. 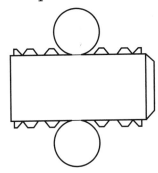	Tape the pattern together. 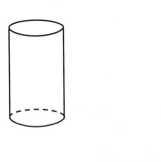

This pattern makes a cylinder. So, the architect used this pattern to build a cylinder.

a.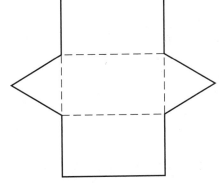

- **What if** the architect wants to build a rectangular prism for a model? How many faces will she draw to make a net for the rectangular prism?

TRY IT

Use the net at the right.

b.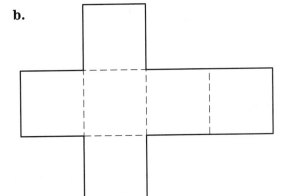

a. Trace the net. Then cut it apart and tape it together. What solid figure did you make?

_____

b. Trace the net. Then cut it apart and tape it together. What solid figure did you make?

_____

CONNECT

Understanding nets can help you draw three-dimensional figures.

ACTIVITY 2 Materials: ruler, triangular dot paper

Rhonda drew a three-dimensional figure that has a length of 4 units and a width of 4 units. The height is 2 units.

STEP 1

On the dot paper, draw a square face that is 4 units long by 4 units wide.

STEP 2

From the vertex, draw 3 vertical line segments that are 2 units high.

STEP 3

Connect 4 units for the length and 4 units for the width to complete the figure.

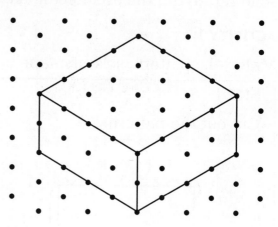

Practice and Problem Solving

For 1-2, use the net.

1. What solid figure would this net make?

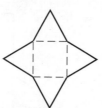

_____

2. What solid figure would this net make?

_____

PACT Test Prep

Identify and draw the solid figure described.

3. I am a solid figure with six congruent square faces. I am 2 units high by 2 units long by 2 units wide. What am I? _____

STEP 1 Draw a square face that is 2 units long by 2 units wide.

STEP 2 From the vertex, draw 3 vertical line segments that are 2 units high.

STEP 3 Connect 2 units for the length and 2 units for the width to complete the figure.

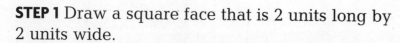

Name _____

PROBLEM SOLVING ON LOCATION

in Myrtle Beach

With more than 90 golf courses, Myrtle Beach and its surrounding area is known as the "Seaside Golf Capital of the World." These golf courses are ranked among the nation's best. One highly acclaimed course is the King's North course at Myrtle Beach National.

1. The second hole at King's North is about 376 yards long. About how many feet is this?

_____

2. The fifteenth hole, one of the longest on the course, is about 487 yards long. Estimate the length of your step or stride. About how many steps would it take you to walk the 487 yards?

_____

3. On the twelfth hole, the tournament tee lies 19 yards behind the regular men's tee. Is this greater than or less than 700 inches? Explain.

_____

4. The tenth hole is 461 yards long. If a golfer is 150 yards from the hole, how far, in feet, has she already hit the ball?

_____

5. A section of the tee area for the first hole is rectangular shaped and has a total area of 54 square yards. If it is 18 ft wide, how long is it?

_____

6. The driving distance from Columbia to Myrtle Beach is about 138 miles. Greenville is about 105 miles farther away. About how far is it to drive from Columbia to Greenville?

_____

BASEBALL AT MYRTLE BEACH

The Myrtle Beach Pelicans is the city's minor league baseball team. Beginning in the Spring of 1999, the Pelicans' games at Coastal Federal Field have become popular with both local residents and tourists visiting Myrtle Beach.

7. The bases on the field form a square which measures 90 feet on each side. What is the area of the field inside the bases?

_____

8. During the first game at Coastal Federal Field, a Pelicans player hit a home run in the fourth inning. How far did he run on one trip around the bases?

_____

9. Coastal Federal Field's rectangular scoreboard measures 48 feet tall and 74 feet wide. What is the area of the scoreboard?

_____

10. Located in left field is a video board that measures 120 inches tall and 156 inches wide. What are these measurements in feet?

_____

11. The outfield wall is lined with 21 billboards. Each billboard is 24 feet long and has an area of 192 square feet. How tall is each billboard?

_____

12. The field is surrounded by a warning track that is 1,308 feet long. How much shorter is this distance than $\frac{1}{4}$-mile?

_____

13. Baseballs used during the Pelicans' games each weigh about 5 ounces. About how much, in pounds, do 16 baseballs weigh?

_____

Name _____

Multiple-Stage Events

 Learn

Salita is going to toss this quarter and spin this pointer. What are the possible outcomes? Which outcomes are equally likely?

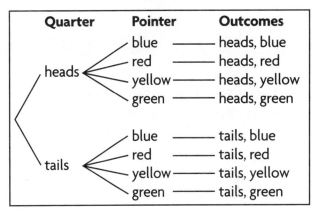

You can use a tree diagram to find all of the possible outcomes.

Since heads and tails are equally likely and blue, red, yellow, and green are equally likely, all of the outcomes are equally likely.

Quarter	Pointer	Outcomes
heads	blue	heads, blue
	red	heads, red
	yellow	heads, yellow
	green	heads, green
tails	blue	tails, blue
	red	tails, red
	yellow	tails, yellow
	green	tails, green

- What is another method you could use to record the possible outcomes of a coin and a spinner?

EXAMPLES Mike and Alicia are playing a game where they toss a coin and spin the pointer shown at the right. Alicia wins if her outcome is heads on the coin and red on the spinner. Mike wins if his outcome is tails on the coin and blue on the spinner. What are the possible outcomes? Are (heads, red) and (tails, blue) equally likely?

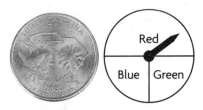

COIN	COLOR		
	Red	**Blue**	**Green**
Heads	heads, red	heads, blue	heads, green
Tails	tails, red	tails, blue	tails, green

Since the red and blue sections are not the same size, red and blue are not equally likely. So, the outcomes (heads, red) and (tails, blue) are not equally likely.

 Check

1. **Explain** how to find the possible outcomes of tossing a coin and spinning a pointer. _____

_____

For 2–3, use the spinner and coin.

2. Find the possible outcomes of tossing the coin and spinning the pointer. Show your work.

_____

3. Are the outcomes all equally likely? Explain.

_____

▶ Practice and Problem Solving

For 4–6, use the spinner and coin.

4. Find the possible outcomes of tossing the coin and spinning the pointer. Show your work.

_____

5. Look at the possible outcomes. Which outcomes are equally likely?

_____

6. Which outcomes are more likely than (heads, yellow) and (tails, yellow)?

_____

PACT Test Prep

7. Team A in Mrs. Wright's class made spinners to decide which state symbols to use in their display. Which is a possible outcome?

   A tree-fruit   **C** tree-flower

   B bird-fruit   **D** flower-flower

TEAM A

8. Team B used spinners to make posters. Record the possible outcomes of spinning the two pointers. Explain the method used and determine whether outcomes are equally likely.

TEAM B

© Harcourt

_____

Name _____

Points on a Grid

▶ Learn

The downtown community of Aiken, South Carolina has a pattern of streets laid out in checkerboard style. You can walk on a tour and see many landmarks and historic sites.

ACTIVITY Take your own tour of Aiken to visit some of the sites.

Materials: coordinate grid, markers

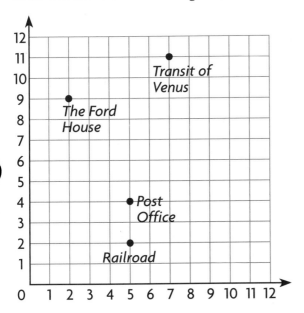

SITE 1 Start at 0. Move right 5 units and then up 2 units. Graph the point and label it "Railroad". Use an ordered pair to describe its location.

_____

SITE 2 Start at the Railroad. Then move left 3 units and up 7 units. Graph the point and label it The Ford House. Use an ordered pair to describe its location. _____

SITE 3 Start at the Ford House. Move up 2 units and then right 5 units. Graph the point and label it Transit of Venus. Use an ordered pair to describe its location. _____

SITE 4 Start at Transit of Venus. Move down 7 units and then left 2 units. Graph the point and label it the Post Office. Use an ordered pair to describe its location. _____

▶ Check

1. **Describe** the location of another site. If you start at The Ford House and move **right** 8 units and then **down** 6 units you arrive at the Nightingale House. Write the ordered pair.

_____

© Harcourt

Unit 9

In Aiken, there are seven elementary schools. Graph a point and write a label for some of these schools. Use an ordered pair to describe its location.

2. Start at 0. Millbrook Elementary is located 3 units right and 3 units up.

_____

3. From Millbrook, North Aiken Elementary is located 1 unit right and 4 units up.

_____

4. From North Aiken, East Aiken Elementary is located 2 units right and 2 units down.

_____

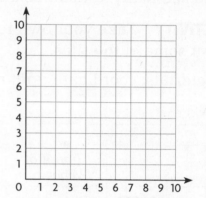

5. A class from East Aiken is on a field trip to the historic Joye Cottage Stable. It was originally used as quarters for horses. From East Aiken, the Stable is located 2 units left and 1 unit down. Write the ordered pair.

_____

6. Graph the ordered pairs (1,4), (4,4), (4,1), and (2,3) on a grid. Then label these points A, B, C, and D. Tell how to move point D so that the points connect to form a square. Name the new ordered pair.

_____

PACT Test Prep

For 7–8, use the grid at the right. Square *ABCD* was moved 6 units down.

7. Describe how point *A* moved. Then write the ordered pair for its new location.

_____

8. Which statement describes how point *B* moved?

   A point *B* moved 5 units over
   B point *B* moved 3 units down
   C point *B* moved 1 unit left and 5 units down
   D point *B* moved 6 units down

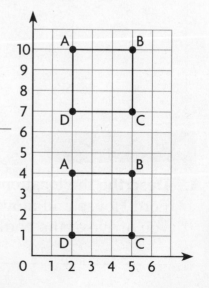

Paths on a Grid

▶ **Learn**

In downtown Columbia, Franco wants to walk from the corner of Main Street and Senate Street to the corner of Sumter Street and Lady Street. He wants to walk the shortest possible path. If he stays on the streets, describe the different paths he can take.

There are three different paths Franco can take. Each path is three blocks long.

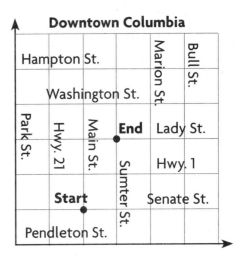

Downtown Columbia

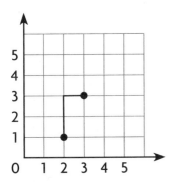 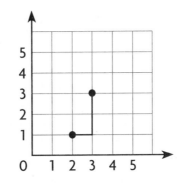 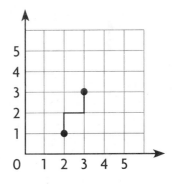

Path 1: 2 blocks up on Main Street and 1 block right on Lady Street.

Path 2: 1 block right on Senate Street and 2 blocks up on Sumter Street.

Path 3: 1 block up on Main Street, 1 block right on Hwy. 1, and 1 block up on Sumter.

EXAMPLE Describe the shortest path you can take from the point at (2,4) to the point at (4,1).

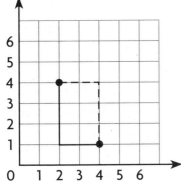

Key:

Path 1 – – – –

Path 2 ———

Path 1: Move 2 units to the right. Move 3 units down.

Path 2: Move 3 units down. Move 2 units to the right.

You can describe other paths to go from (2,4) to (4,1), but they will be longer paths.

© Harcourt

For 1–2, use the map at the right.

Downtown Columbia

Hampton St.			Marion St.	Bull St.
	Washington St.			
Park St.	Hwy. 21	Main St.		Lady St.
			Sumter St.	Hwy. 1
				Senate St.
	Pendleton St.			

1. Describe Franco's path if he walks along Senate Street from Park to Marion.

_____

2. Helena takes the shortest possible path from the corner of Park and Washington to Main and Lady. Describe one possible path.

_____

► **Practice and Problem Solving**

For 3–6, use the map above.

3. Tommy walks the shortest path from Sumter and Pendleton to Main and Lady. Describe one possible path.

_____

4. Look at Problem 3. Find the shortest path that Tommy can take if he makes two turns.

_____

PACT Test Prep

5. Greg walks 3 blocks right from the corner of Main and Washington. Where does this path take him?

   A Marion and Washington
   B Bull and Washington
   C Bull and Hampton
   D Bull and Lady

6. Give directions to tell someone how to get from Bull and Hampton to Main and Hwy. 1 by taking the shortest path.

   F 3 blocks down, 3 blocks left
   G 1 block down, 3 blocks left, 1 block down
   H 3 blocks down, 4 blocks left
   J 3 blocks left, 4 blocks down

7. Use the grid at the right. Draw two possible paths from (2,5) to (5,2). Describe each path that you drew.

_____

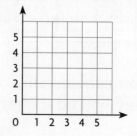

PROBLEM SOLVING ON LOCATION

in Darlington County

Darlington County is located in the northeastern part of the state. The Darlington Raceway and Kalmia Gardens are among its most popular attractions. Part of Darlington County's appeal is its moderate temperatures. This table shows the average monthly temperatures in Darlington.

For 1–5, use the table.

1. Which month is the coldest? Justify your answer.

_____

2. What is the change in average low temperature from July to December?

_____

3. Find the median of the average monthly high temperatures.

_____

4. The average low temperatures for February to May can be written as the equation $y = 4x - 6$, where x represents the number of the month. For example, February = 2 and March = 3. Graph the equation for the 4 months.

_____

5. The record low temperature for March is 16 degrees below its average low temperature. Find the record low temperature for March.

_____

AVERAGE TEMPERATURE IN DARLINGTON, SC (IN °C TO THE NEAREST DEGREE)		
Month	**Low**	**High**
January	0	13
February	2	15
March	6	20
April	10	25
May	14	29
June	19	32
July	21	33
August	20	32
September	17	30
October	10	25
November	6	20
December	2	15

© Harcourt

KALMIA GARDENS AT COKER COLLEGE

Kalmia Gardens is located in Darlington County in Hartsville, South Carolina, and has been open to the public since 1935. A unique 60-foot drop in elevation within the Gardens provides a variety of plant and animal life enjoyed by visitors year-round.

6. Anna says the drop in elevation within Kalmia Gardens is 180 yards. Describe and correct her error.

_____

7. Robert lives 8.9 km from Kalmia Gardens. If Melanie lives 5 km less than twice that distance from the Gardens, what is the distance from Melanie's house to Kalmia Gardens?_____

Jessica planted some of the flowers she had seen at Kalmia Gardens. She used a coordinate grid to map where she planted each type of flower.

For 8–11, use the grid.

8. Write an ordered pair for the location where each flower was planted.

_____

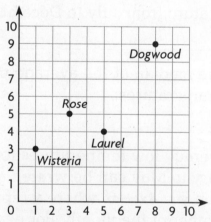

9. Name the plant that is located 2 units down and 2 units left from the rose._____

10. Name the plant that is located 3 units right and 5 units up from the laurel. _____

11. Name the plant that is located 1 unit down and 2 units right from the rose. _____

12. Carl is choosing a project about his visit to Kalmia Gardens. He can make a card, a calendar, a postcard, or a poster. He can feature trees, flowering shrubs, or herbs. In how many ways can he do the project? _____

13. Suppose Jessica wanted to plant a beech tree at (8,10). Describe how you would graph this point on a coordinate grid.

_____

_____

Choose the best answer.

1. Mrs. Bennett distributed 326 pencils to the students each month at the library. Which is the most reasonable estimate for the number of pencils used in one year? NO III.C.2

   Ⓐ 4,000   C 9,000
   B 6,000   D 30,000

2. The pastry chef used 56 pounds of flour at the restaurant each week. If she used the same amount every night, write an equation to find n, or the number of pounds of flour she used per night. NO II.C.1

   F $n \times 56 = 56$ Ⓗ $n \times 7 = 56$
   G $56 \div 8 = n$ J $8 \times n = 56$

3. If you know that $35 \times 40 = 1,400$, how can you find 35×80? NO II.B.1

   A Write a zero after the product.
   B Find half of 1,400.
   C Write the same product.
   Ⓓ Double 1,400.

4. What is 3,077,590 rounded to the nearest hundred thousand? NO III.C.1

   F 3,000,000
   Ⓖ 3,100,000
   H 3,200,000
   J 4,000,000

5. Rachel bought a pair of skates for $22. Her sister bought a pair for $47. Which estimated difference is closer to the exact answer? NO III.C.4

   A $2 B $15 Ⓒ $30 D $45

6. In the next six days, Lexi wants to run a total of 18 laps. She wants to run the same number of laps each day. Which equation can she use to solve the problem? A II.C.1

   F $18 + 6 = n$ H $18 - 6 = n$
   Ⓖ $6 \times n = 18$ J $6 \times 18 = n$

For 7–8, use the bar graph.

Miles Run in Marathon

7. About how many more miles did Sophia run than Victor? DAP I.C.2

   A 10 mi   C 4 mi
   B 6 mi   Ⓓ 2 mi

8. Which is the best estimate of the total number of miles the marathoners ran? DAP I.C.2

   F 80 mi   H 65 mi
   Ⓖ 70 mi   J 50 mi

For 9–10, use the table.

FOURTH GRADE FIELD TRIP SCHEDULE	
Bus Departs	7:30 A.M.
Enter Park	9:00 A.M.
Lunch	12:30 P.M.
Leave Park	4:45 P.M.
Arrive at School	6:15 P.M.

9. The fourth-grade class went to a water park for a field trip. This is a schedule for the trip. The class had to leave the park 1 hour 15 minutes early due to a rainstorm. At what time did the class leave the park? M II.B.3

   A 2:15 P.M.   Ⓒ 3:30 P.M.
   B 2:30 P.M.   D 3:45 P.M.

10. The bus was early arriving back at school. Use the clock below to read the time the students arrived back at school. M II.B.4

   F 25 minutes after four
   G 25 minutes after six
   Ⓗ 20 minutes after five
   J 25 minutes to five

11. What is 7,905,410 written in expanded form? NO I.B.2

   A 700,000 + 900,000 + 5,000 + 400 + 10
   B 7,000,000 + 9,000 + 5,000 + 400 + 10
   C 7,000,000 + 900,000 + 500 + 400 + 10
   Ⓓ 7,000,000 + 900,000 + 5,000 + 400 + 10

12. Sarah is participating in a dance competition. She will dance in five different routines for a total of 1 hour 30 minutes. If each routine takes the same amount of time, how much time will each take? M I.C.2

   F 8 min   H 12 min
   G 10 min   Ⓙ 18 min

13. Rebecca wants to calculate how much it would cost to buy 17 tickets to a soccer game at $5 per ticket. Which expression shows a way to calculate 17×5 mentally? NO II.D.2

   A $(7 \times \$5) + 10$
   Ⓑ $(10 \times \$5) + (7 \times \$5)$
   C $10 \times 7 \times \$5$
   D $10 \times \$5 + 7$

For 14–15, use the line graph.

Alison's Bank Account

14. Which best describes the change in Alison's bank account between weeks 4 and 5? DAP III.A.1

   F Alison put more money into her bank account.
   Ⓖ Alison took money out of her bank account.
   H Alison did not go to the bank.
   J Alison gave money to the bank teller.

   A IV.B.1
15. What do you notice about the change in the number of dollars saved between weeks 1 and 5?

   A The number of dollars saved decreases, then increases.
   B The number of dollars saved increases.
   Ⓒ The number of dollars saved increases, then decreases.
   D The number of dollars saved decreases.

16. How many periods are in a 7 digit number? NO I.A.1

   F 4 Ⓖ 3 H 2 J 1

17. The book fair committee purchased 75 pocket folders at $1.25 each. Choose the equation to find the total cost. A II.C.1

   A $c \times 75 = \$1.25$
   Ⓑ $c = 75 \times \$1.25$
   C $c = 75 - \$1.25$
   D $c \times \$1.25 = 75$

For 18, use the table.

FAVORITE INSTRUMENTS	
clarinet	8
flute	5
trumpet	5
oboe	3
percussion	10

18. Justin took a vote in his school's band of students' favorite instruments. Which instrument represents the mode? DAP II.B.1

   Ⓕ percussion H clarinet
   G flute   J trumpet

19. What is the value of the digit 4 in the number 5,435,650,000? NO I.A.1

   A 4,000   C 40,000
   Ⓑ 400,000,000 D 4,000,000,000

20. The heartbeat of an average dog weighing 30 pounds is about 100 beats per minute. At that rate, how many times would a dog's heart beat in 20 minutes? NO III.A.1

   F 200,000   Ⓗ 2,000
   G 20,000   J 200

---

21. The Company Theater holds 462 people. Over five days, there will be seven performances of an amateur play. If each play is sold out, how many tickets have been sold? Tell which place-value positions need to be regrouped. [1 point] NO III.B.2

   3,234 people; tens, hundreds

22. What method would you use—paper and pencil or mental math—to compute 830×50? Explain. [1 point] NO III.F.1

   Possible answer: mental math; $800 \times 50 = 40,000$; $30 \times 50 = 1,500$;
   $40,000 + 1,500 = 41,500$

23. Ross records one science test score every week for 8 weeks. What kind of graph would be best to display this data? Explain. [1 point] DAP I.D.2

   Possible answer: line graph; the line graph shows the change in
   Ross' test scores over a period of time.

24. The Emperor penguin is the largest penguin that weighs about 80 pounds. How many pounds does a pack of 18 penguins weigh? Describe how you can make a model using grid paper to find the product 80×18. Tell what the product is. [1 point] NO III.B.1

   Outline a rectangle that is 80 units long and 18 units wide.
   Break apart the model so you can show $80 = 40 + 40$ and $18 = 9 + 9$;
   multiply; then add the partial products: $360 + 360 + 360 + 360 = 1,440$;
   about 1,440 lb

25. When marine turtles lay eggs, they usually lay them in batches or clutches of six. Each clutch contains about 100 eggs. If there are 4 female marine turtles, about how many eggs will they lay? [1 point] NO III.A.1

   about 2,400 eggs; $4 \times 6 \times 100 = 2,400$

© Harcourt

For 26, use the table.

INPUT	p	3	4	5	6	7
OUTPUT	r	$18	$24	■	$36	$42

26. At Wild Run State Park, it costs $18 to rent a canoe for 3 hours. Brian and Alan want to go canoeing for 5 hours. How much will they have to pay? Find a rule for the table and name the missing number. Then write the equation. [2 points] A I.A.1, I.B.1, I.B.2, II.B.1, II.C.1

$30; multiply by 6; $r = p \times \$6$

For 27, use the bar graphs.

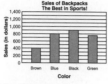

27. Ryan's job at Action Sports Manufacturing Company is to analyze the sales of backpacks at two leading sports stores. The graphs show the data he collected. Compare the shapes of the graphs. What conclusions can you draw from the sales data? Describe how Ryan can use the results. [3 points] DAP II.A.1

Possible answers: The number of black backpacks sold is the greatest at both stores; the number of backpacks sold at The Best in Sports! is about the same for each color; Ryan can tell his supervisor that red backpacks were not in the top four of backpacks sold at The Best in Sports!; brown backpacks were in the top four; Ryan can also report that they should manufacture more black backpacks.

28. Mr. Sheridan's class took a survey of fourth grade students to find their favorite types of movies. The table below has the survey's results. Make a bar graph that shows the results of the survey. Be sure to
- give your graph a title
- use a scale
- tell what each bar represents
- write two statements that compare the data on types of movies. [3 points] DAP I.C.1, I.C.2

TYPE OF MOVIE	NUMBER OF STUDENTS
Cartoon	14
Comedy	21
Science Fiction	9
Adventure	16

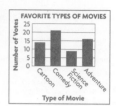

Possible answers: The results of the survey show that most of the students like a comedy type of movie or an adventure. The fewest number of students like a science fiction movie.

Item	Standard
1	NO III.C.2
2	NO II.C.1
3	NO II.B.1
4	NO III.C.1
5	NO III.C.4
6	A II.C.1
7	DAP I.C.2
8	DAP I.C.2
9	M II.B.3
10	M II.B.4
11	NO I.B.2
12	M I.C.2
13	NO II.D.2
14	DAP III.A.1
15	A IV.B.1

Item	Standard
16	NO I.A.1
17	A II.C.1
18	DAP II.B.1
19	NO I.A.1
20	NO III.A.1
21	NO III.B.2
22	NO III.F.1
23	DAP I.D.2
24	NO III.B.1
25	NO III.A.1
26	A I.A.1, I.B.1, I.B.2, II.B.1, II.C.1
27	DAP II.A.1
28	DAP I.C.1, I.C.2

Scoring Rubric

2 Demonstrates a complete understanding of the problem and chooses an appropriate strategy to determine the solution

1 Demonstrates a partial understanding of the problem and chooses a strategy that does not lead to a complete and accurate solution

0 Demonstrates little understanding of the problem and shows little evidence of using any strategy to determine a solution

To score a 3-point question, differentiate between levels of partial understanding demonstrated by the student. A "2" response communicates a general understanding of the problem and process needed to solve it, but falls short of being a thoroughly correct response. A "1" response communicates some understanding, but merely begins to approach a solution.

© Harcourt

Choose the best answer.

1. Betty left for school at 8:23 A.M. She arrived at 9:00 A.M. How many minutes did it take her to get to school? M II.B.3

   A 23 min C 47 min
   (B) 37 min D 53 min

2. There are 3 quarts of punch. How many 1-cup servings will 3 quarts make? M I.C.1

   | 1 quart = 2 pints |
   | 1 pint = 2 cups |

   F 3 cups H 6 cups
   G 4 cups (J) 12 cups

For 3, use the solid figures.

3. Which of the following statements is true? G I.A.1

   A A cube has more faces than a rectangular prism.
   (B) A cube has the same number of faces as a rectangular prism.
   C A cube has fewer faces than a rectangular prism.
   D The number of faces depends on the size of the cube or rectangular prism.

4. There are 94 students separated into equal groups and 4 students left over. Each group has more than 15, but fewer than 20 members. How many groups are there? NO III.B.1

   F 3 (H) 5
   G 4 J 6

5. What are the factors of 49? NO I.G.1

   (A) 1, 7, 49
   B 1, 7, 9, 49
   C 1, 3, 7, 9, 49
   D 2, 49, 98

For 6, use the spinner.

6. The spinner has four equally likely outcomes. What is the probability of the pointer stopping on 4? DAP IV.B.1

   F $\frac{1}{8}$ (H) $\frac{1}{4}$
   G $\frac{1}{6}$ J $\frac{1}{2}$

For 7, use the drawing.

door

7. Jan was facing the door of her classroom. She turned to her right until her back was facing the door. How many degrees did she turn? M I.A.2

   A 45° (C) 180°
   B 90° D 270°

8. In science class, four of the students have plants with heights 4.25 cm, 5.36 cm, 4.52 cm, and 5.63 cm. Which statement is true? NO I.A.2

   F 4.25 cm is greater than 5.36 cm
   (G) 4.25 cm is less than 5.36 cm
   H 5.63 cm is less than 5.36 cm
   J 4.25 cm is greater than 4.52 cm

   Go On

For 9, use the model.

9. What decimal and fraction are shown by the model? NO I.E.3

   A 25; $\frac{25}{1}$ (C) 0.25; $\frac{25}{100}$
   B 2.5; $2\frac{5}{10}$ D 0.025; $\frac{25}{100}$

10. Edward lives 3.9 km from the library. Sandy lives 500 m more than twice that distance from the library. How far from the library does Sandy live? M I.C.1

    (F) 8.3 km H 7.3 km
    G 7.8 km J 4.5 km

For 11, use the figure.

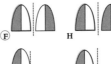

11. What is the area? M II.D.1

    A 36 sq cm C 45 sq cm
    B 40 sq cm (D) 48 sq cm

12. Which of the following shows a flip? G III.B.2

    (F)   H

    G   J

13. Matt and Jennie are playing a game called *Name the Pattern Rule*. Matt names these numbers: 2.045, 6.135, 18.405, 55.215. Jennie names a rule for the pattern. Which rule does she name? A I.A.1

    A Add 4.09. C Multiply by 2.
    B Add 12.27. (D) Multiply by 3.

For 14, use the drawing.

14. What terms describe this triangle by sides and by the measure of its angles? G I.B.1

    F scalene, obtuse
    (G) equilateral, acute
    H isosceles, right
    J equilateral, obtuse

For 15, use the figure.

15. What is the best estimate of the area of the polygon drawn on the grid? M II.A.1

    (A) 28 sq units C 40 sq units
    B 32 sq units D 62 sq units

16. Which is a solid figure with 3 rectangular faces and 2 triangular faces? G IV.B.1

    F cube H rectangular prism
    G cylinder (J) triangular prism

    Go On

For 17, use the model.

17. Compare the fractions. Which is true? NO I.D.2

    A $\frac{4}{5} < \frac{2}{3}$ C $\frac{2}{3} > \frac{4}{5}$
    (B) $\frac{4}{5} > \frac{2}{3}$ D $\frac{2}{3} = \frac{4}{5}$

For 18, answer the question and tell how to interpret the remainder.

18. A 61-inch piece of ribbon needs to be cut into 8-inch lengths. How many 8-inch lengths will there be? NO II.A.1

    F 8; increase the quotient by 1
    (G) 7; drop the remainder
    H 7; use the remainder as the answer
    J 5; use the remainder as the answer

19. Charles read the temperature at 7 P.M. at 25°F. During the night, the low temperature was ⁻5°F. By how many degrees did the temperature drop? M II.B.5

    A 40° B 35° (C) 30° D 20°

20. Brittany's favorite TV program comes on at 4:00 P.M. John's favorite program is on $\frac{1}{2}$ hour later and lasts $1\frac{1}{2}$ hours. At what time does John's favorite program end? M II.B.3

    F 4:30 P.M. H 5:30 P.M.
    G 5:00 P.M. (J) 6:00 P.M.

21. Sheila has a number cube that is labeled 2, 2, 4, 4, 6, and 6. What is the probability that Shelia will toss a 2 or a 4? DAP IV.C.1

    A $\frac{5}{6}$ (B) $\frac{4}{6}$ C $\frac{3}{6}$ D $\frac{2}{6}$

For 22, use the model.

22. What is the sum of 1.77 + 1.44? NO III.E.1

    (F) 3.21 H 2.21
    G 3.11 J 2.11

For 23–24, use the grid.

(grid with triangle)

23. What is the ordered pair for the corner of the triangle that is **not** a corner of the square? G II.B.2

    A (2,1) B (2,5) (C) (4,8) D (6,5)

24. Start at (4,9). Which describes how to move to (6,1)? G IV.B.2

    F 4 units down, 2 units left
    (G) 2 units right, 8 units down
    H 8 units down, 2 units left
    J 4 units down, 2 units right

    Go On

---

25. Eric has 15 tomato plants. He wants the plants to be 3 feet apart and also 3 feet from the garden fence. Draw a diagram of his garden. What is a reasonable area for Eric's garden? Explain. [2 points] M I.A.3

    Check students' drawings.

    216 sq ft; Possible answer: Find the length by using each 3-ft space between plants and next to the fence, 6 × 3 = 18, or 18 ft; find the width, 4 × 3 = 12, or 12 ft; multiply length by width to find the area, 18 × 12 = 216, or 216 sq ft.

For 26, use the drawing at the right.

26. Janie drew this figure. How can she draw this shape so that it shows a slide? How can she draw the original shape so that it shows a flip? Explain. [1 point] G III.B.1

Slide

Flip

    Possible answer: A slide moves a figure to a new position; a flip is a movement of a figure to a new position by flipping the figure over a line.

27. The Ruiz family visited the Columbia Museum of Art in Columbia, South Carolina. They saw points, lines, line segments, rays, and angles in many of the paintings. Draw a picture to show an example of each term. Label your drawings with the terms that describe them. [1 point] G IV.A.1

    Check students' drawings. Drawings should include labels for all the geometric terms.

    Go On

© Harcourt

For 28, use the figure.

Possible drawings:

28. Divide the figure into 4 equal parts. Find the area of the whole figure. Then shade $\frac{1}{2}$ of the figure. Explain how you know that the parts you shaded are $\frac{1}{2}$ of the whole figure. [3 points] G I.C.1

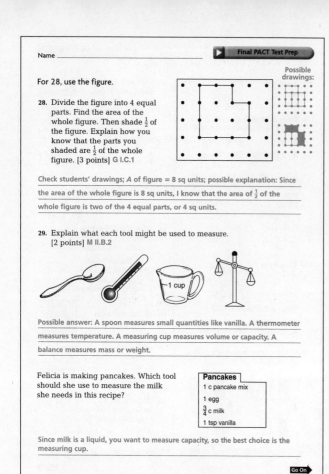

Check students' drawings; *A* of figure = 8 sq units; possible explanation: Since the area of the whole figure is 8 sq units, I know that the area of $\frac{1}{2}$ of the whole figure is two of the 4 equal parts, or 4 sq units.

29. Explain what each tool might be used to measure. [2 points] M II.B.2

Possible answer: A spoon measures small quantities like vanilla. A thermometer measures temperature. A measuring cup measures volume or capacity. A balance measures mass or weight.

Felicia is making pancakes. Which tool should she use to measure the milk she needs in this recipe?

Pancakes
1 c pancake mix
1 egg
$\frac{3}{4}$ c milk
1 tsp vanilla

Since milk is a liquid, you want to measure capacity, so the best choice is the measuring cup.

---

30. Suppose Mary spins the pointer on each spinner. Make a tree diagram to list all the possible outcomes for the event. [3 points] DAP IV.A.1, IV.B.2

green / red — **Spinner 1**

1 2 / 4 3 — **Spinner 2**

Tree diagram
Check students' diagrams. There are 8 possible outcomes.

Spinner 1	Spinner 2	Outcomes
red	1	red-1
	2	red-2
	3	red-3
	4	red-4
green	1	green-1
	2	green-2
	3	green-3
	4	green-4

Which outcomes are equally likely? Which outcomes are more likely? Explain your answer.
red-1, red-2, red-3, red-4 are equally likely; green-1, green-2, green-3, green-4 are also equally likely; red-1, red-2, red-3, red-4 are more likely than green-1, green-2, green-3, green-4; Possible explanation: Since the green and red sections on the spinner are not the same size, green and red are not equally likely. So, the outcomes (green-1) and (red-2) are not equally likely.

Item	Standard	Item	Standard
1	M II.B.3	16	G IV.B.1
2	M I.C.1	17	NO I.D.2
3	G I.A.1	18	NO II.A.1
4	NO III.B.1	19	M II.B.5
5	NO I.G.1	20	M II.B.3
6	DAP IV.B.1	21	DAP IV.C.1
7	M I.A.2	22	NO III.E.1
8	NO I.A.2	23	G II.B.2
9	NO I.E.3	24	G IV.B.2
10	M I.C.1	25	M I.A.3
11	M II.D.1	26	G III.B.1
12	G III.B.2	27	G IV.A.1
13	A I.A.1	28	G I.C.1
14	G I.B.1	29	M II.B.2
15	M II.A.1	30	DAP IV.A.1, IV.B.2

Scoring Rubric

2 Demonstrates a complete understanding of the problem and chooses an appropriate strategy to determine the solution

1 Demonstrates a partial understanding of the problem and chooses a strategy that does not lead to a complete and accurate solution

0 Demonstrates little understanding of the problem and shows little evidence of using any strategy to determine a solution

To score a 3-point question, differentiate between levels of partial understanding demonstrated by the student. A "2" response communicates a general understanding of the problem and process needed to solve it, but falls short of being a thoroughly correct response. A "1" response communicates some understanding, but merely begins to approach a solution.

© Harcourt

Page 1 (worksheet top-left)

Name _____

Place Value Through Billions

 Learn

In 1998, the value of products produced and services performed in South Carolina was $100,350,000,000. If you counted one dollar every second, it would take almost 3,180 years to reach $100,350,000,000!

> VOCABULARY
> billions

A place-value chart can help you read and write greater numbers. The period to the left of *millions* is **billions**.

BILLIONS			**MILLIONS**			**THOUSANDS**			**ONES**		
Hundreds	Tens	Ones	Hundreds	Tens	Ones	Hundreds	Tens	Ones	Hundreds	Tens	Ones
1	0	0,	3	5	0,	0	0	0,	0	0	0

Standard Form: 100,350,000,000

Word Form: one hundred billion, three hundred fifty million

Expanded Form: 100,000,000,000 + 300,000,000 + 50,000,000

EXAMPLES

Standard Form	Word Form	Expanded Form
A 6,033,200,000	six billion, thirty-three million, two hundred thousand	6,000,000,000 + 30,000,000 + 3,000,000 + 200,000
B 50,150,000,905	fifty billion, one hundred fifty million, nine hundred five	50,000,000,000 + 100,000,000 + 50,000,000 + 900 + 5

 Check

1. **Tell** how many digits are in the number ten billion. _11 digits_

Write the value of the digit 8 in each number.

2. 38,320,400,090
   8,000,000,000

3. 846,390,000,290
   800,000,000,000

4. 9,822,453,777
   800,000,000

Page 2 (worksheet top-right)

Write the value of the underlined digit.

5. 534,888,245,454
   30,000,000,000

6. 5,800,000,050
   5,000,000,000

7. 65,875,500,000
   800,000,000

Write each number in standard form.

8. three hundred billion, four hundred thousand
   300,000,400,000

9. one billion, fifty million
   1,050,000,000

Write each number in word form.

10. 40,000,000,000
    forty billion

11. 6,500,000,000
    six billion, five hundred million

12. Complete. Explain how place value and period names help you read and write numbers.
    5,002,500,000 = 5 _billion_ + 2 _million_ + 5 hundred thousand
    Possible answer: The position of each digit in the period determines its value.

PACT Test Prep

For 13–14, use the table.

13. What is the value of the digit 1 in the amount for transportation? NO I.A.1; DAP I.C.2
    (A) $1,000,000,000   C $100,000
    B $1,000,000   D $1,000

14. If Health receives one hundred million dollars more in next year's budget, what will its budget be? Write this amount in standard form. NO I.B.2; DAP I.C.2
    $5,294,564,000

SOUTH CAROLINA STATE BUDGET 2000–2001 (rounded to the nearest thousand)	
Department	Amount
Health	$ 5,194,564,000
K-12 Education	$ 2,805,620,000
Higher Education	$ 2,263,941,000
Other	$ 1,847,265,000
Transportation	$ 1,027,802,000
Public Safety	$ 763,907,000
Total	**$13,889,210,000**

Lesson 1.4A

Objective To read, write, and identify the place value of whole numbers through billions

South Carolina Standards Number and Operations I.A.1 Explain the place value structure of whole numbers including periods (thousands, millions, billions, etc.); I.B.2 Write whole numbers in standard form, in expanded form, and in words; Data Analysis and Probability I.C.2 Read and interpret information from tables, line graphs, and bar graphs.

Vocabulary billions

Introduce

WHY LEARN THIS? Greater numbers, such as the populations of countries or the sales data of corporations, will be easier to read and understand.

Review place value with students. Display a place-value chart to hundred millions. Have students read the number 23,070,100 aloud. Ask them to identify each digit's place-value position on the chart. Invite students to point to the place-value position that changes when you add 100,000,000 to the number. **What is the new number?** 123,070,100 **Add 10,000,000 to the original number. What is the new number?** 33,070,100

Using the Pages

Guide students through Exercises 1–4. Point out that one billion has one more period than one million. Ask: **Since there are 1,000 millions in 1 billion, how many millions are in 10 billion? in 100 billion?** 10,000; 100,000

Assess

Summarize the lesson by having students: **Write each value of the digit 7 in the number 377,174,752,000. Start with the greatest place-value position.** 70,000,000,000; 7,000,000,000; 70,000,000; 700,000

PACT Test Prep

Item	Standard
13	NO I.A.1; DAP I.C.2
14	NO I.B.2; DAP I.C.2

PROBLEM SOLVING ON LOCATION
in South Carolina's Cities

In 1786, the capital of South Carolina was moved from the coastal city of Charleston to a more central location. The capital was named Columbia. The population of Columbia in 1786 was about 1,000. The table shows the 2000 population of Columbia and nine other South Carolina cities.

For 1–6, use the table.

1. Which South Carolina city has the greatest population?
   Columbia

2. What is the population of Charleston rounded to the nearest ten thousand?
   100,000

3. Which city has the greater population—Mount Pleasant or Sumter? Explain.
   Mount Pleasant; 47,609 > 39,643

4. Which cities have a population greater than 50,000 people?
   Charleston, Columbia, Greenville, North Charleston

5. Estimate the number of people who live in the three largest cities. Name the cities.
   about 300,000 people, Columbia, Charleston, North Charleston

6. About how many more people live in Columbia in 2000 than in 1786?
   about 115,000 more people

| TEN LARGEST CITIES IN SOUTH CAROLINA ||
City	Population (2000)
Charleston	96,650
Columbia	116,278
Florence	30,248
Greenville	56,002
Hilton Head Island	33,862
Mount Pleasant	47,609
North Charleston	79,641
Rock Hill	49,765
Spartanburg	39,673
Sumter	39,643

South Carolina has grown rapidly in the last 50 years, and currently ranks 26th in population among the 50 states. The table below shows South Carolina's increase in population since 1950.

For 7–15, use the table.

7. Round the state population for 2000 to the nearest million.
   4,000,000

8. Describe how the size of the population changed between the years 1950 and 2000.
   Possible answer: The size of the population doubled.

9. If the population data for 1980 and 1990 are rounded to the nearest million, are the rounded numbers the same? Explain.
   Yes; both numbers round to 3,000,000.

10. Was there a greater population growth between 1980 and 1990 or between 1990 and 2000?
    between 1990 and 2000; 525,309 > 364,883

11. What is the value of the digit 9 in the population for 1970?
    90,000

12. What number is three hundred fifty-six thousand, four hundred nine greater than the state population in 2000?
    4,368,421

13. How much greater is the 1980 population than the population for 1970? 1960?
    531,304 greater; 739,226 greater

14. How can you write South Carolina's population for 2000 in word form?
    four million, twelve thousand, twelve

15. Write South Carolina's population for 1990 in expanded form.
    3,000,000 + 400,000 + 80,000 + 6,000 + 700 + 3

| SOUTH CAROLINA STATE POPULATION ||
Year	Population
1950	2,117,027
1960	2,382,594
1970	2,590,516
1980	3,121,820
1990	3,486,703
2000	4,012,012

Unit 1 Problem Solving
On Location in South Carolina's Cities

Purpose To provide additional practice for concepts and skills from Chapters 1–4

South Carolina Standards Number and Operations I.A.1 Explain the place value structure of whole numbers including periods (thousands, millions, billions, etc.); I.B.2 Write whole numbers in standard form, in expanded form, and in words; III.C.1 Round whole numbers to the nearest ten thousand, hundred thousand, and million; Data Analysis and Probability I.C.2 Read and interpret information from tables, line graphs, and bar graphs.

Using the Pages

Reinforce place value skills by asking: **How do you use place-value to answer Exercise 1?** Possible answer: You can use place value to compare numbers in the table. Columbia is the only city whose population has a digit in the hundred thousands place.

Have students look at Exercise 4. Ask: **Which place-value position do you look at to determine if the population of a city is greater than 50,000?** Possible answer: I look at the ten thousands place, except for Columbia; for Columbia, I look at the hundred thousands place.

Have students look at Exercise 8. Demonstrate the rules of rounding. Then ask: **How could you use rounding to find the answer?** 2,117,027 rounded to the nearest million is 2,000,000; 4,012,012 rounded to the nearest million is 4,000,000; 4,000,000 is twice as great, so the size doubled.

Direct students' attention to Exercise 12. Ask: **What operation do you use to solve this problem? Explain.** Addition; the word *greater* and the relationship of the numbers indicates joining groups of different sizes.

Ask students to look at Exercise 13. **Why do you use subtraction to solve this problem?** Possible answer: The words, *how much greater*, indicate finding the difference.

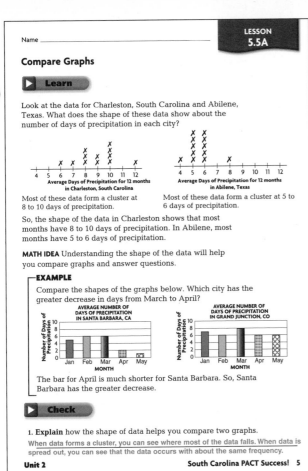

Compare Graphs

▶ **Learn**

Look at the data for Charleston, South Carolina and Abilene, Texas. What does the shape of these data show about the number of days of precipitation in each city?

Average Days of Precipitation for 12 months in Charleston, South Carolina

Average Days of Precipitation for 12 months in Abilene, Texas

Most of these data form a cluster at 8 to 10 days of precipitation.

Most of these data form a cluster at 5 to 6 days of precipitation.

So, the shape of the data in Charleston shows that most months have 8 to 10 days of precipitation. In Abilene, most months have 5 to 6 days of precipitation.

MATH IDEA Understanding the shape of the data will help you compare graphs and answer questions.

EXAMPLE

Compare the shapes of the graphs below. Which city has the greater decrease in days from March to April?

AVERAGE NUMBER OF DAYS OF PRECIPITATION IN SANTA BARBARA, CA

AVERAGE NUMBER OF DAYS OF PRECIPITATION IN GRAND JUNCTION, CO

The bar for April is much shorter for Santa Barbara. So, Santa Barbara has the greater decrease.

▶ **Check**

1. **Explain** how the shape of data helps you compare two graphs.

When data forms a cluster, you can see where most of the data falls. When data is spread out, you can see that the data occurs with about the same frequency.

Unit 2   **South Carolina PACT Success!**   5

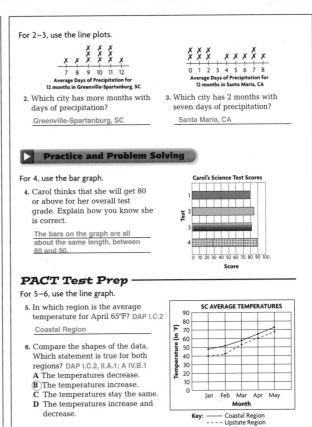

For 2–3, use the line plots.

Average Days of Precipitation for 12 months in Greenville-Spartanburg, SC

Average Days of Precipitation for 12 months in Santa Maria, CA

2. Which city has more months with days of precipitation?

Greenville-Spartanburg, SC

3. Which city has 2 months with seven days of precipitation?

Santa Maria, CA

▶ **Practice and Problem Solving**

For 4, use the bar graph.

4. Carol thinks that she will get 80 or above for her overall test grade. Explain how you know she is correct.

The bars on the graph are all about the same length, between 80 and 90.

Carol's Science Test Scores

PACT Test Prep

For 5–6, use the line graph.

5. In which region is the average temperature for April 65°F? DAP I.C.2

Coastal Region

6. Compare the shapes of the data. Which statement is true for both regions? DAP I.C.2, II.A.1; A IV.B.1
A The temperatures decrease.
B The temperatures increase.
C The temperatures stay the same.
D The temperatures increase and decrease.

SC AVERAGE TEMPERATURES

Key: ——— Coastal Region
       - - - - Upstate Region

6   **South Carolina PACT Success!**   Unit 2

Lesson 5.5A

Objective To compare the shapes of two graphs

South Carolina Standards Algebra IV.B.1 Describe changes over time as increasing, decreasing, and varying using charts and graphs; Data Analysis and Probability I.C.2 Read and interpret information from tables, line graphs, and bar graphs; II.A.1 Compare the shapes of graphs of two different numerical data sets that address the same question for different populations.

Introduce

WHY LEARN THIS? The shape of a graph indicates if data increases, decreases, or stays the same. For example, you can read graphs to compare changes in weather or to compare the speed of sound as it passes through different materials.

Clarify for students the parts of graphs and how to read graphs. Then encourage class discussion about the importance of making comparisons of graphs.

Using the Pages

Guide students through the Check exercises. Point out how the data is shaped on each line plot. Ask: **How do the shapes of the two line plots compare?** Possible answer: The data for Greenville-Spartanburg form a cluster; the data for Santa Maria are spread out.

For Exercise 4, elicit from students that the bars on this graph are all about the same length, so the data are spread out.

Assess

Summarize the lesson by asking students to: **Describe the shape of data when it is clustered on a line plot. Explain what *clustered* means.**

Possible answer: The shape of the data shows X's clustered together; most of the data points are close or gathered together.

PACT Test Prep

Item	Standard
5	DAP I.C.2
6	DAP I.C.2, II.A.1; A IV.B.1

PROBLEM SOLVING ON LOCATION
in South Carolina's Blue Ridge Mountains

The Blue Ridge Mountains extend from northern Georgia to Pennsylvania and cover parts of six states. The name of the mountain range comes from the blue tone of the evergreen trees when they are seen from a distance.

At 3,560 feet, Sassafras Mountain is South Carolina's highest point.

For 1–5, use the table.

1. Which graph would better display the type of data in the table—a line graph or a bar graph? Why?
   A bar graph; it compares data about groups.

2. Order the mountain heights from highest to lowest.
   6,684; 5,964; 5,729; 4,784; 4,049; 3,560

3. Suppose a hiking trip on Sassafras Mountain is two days long. How many hours is this?
   48 hr

4. How many feet higher is Mount Mitchell than Mount Rogers?
   955 ft higher

5. Which two mountains, when their heights are rounded to the nearest hundred, have a difference of 2,000 feet?
   Grandfather and Hawksbill; 6,000 − 4,000 = 2,000

PEAK HEIGHTS BLUE RIDGE MOUNTAINS	
Mountain	**Height (in feet)**
Sassafras	3,560
Mount Mitchell	6,684
Grandfather	5,964
Mount Rogers	5,729
Hawksbill	4,049
Brasstown Bald	4,784

UPCOUNTRY WATERFALLS

In places, the steep Blue Ridge Mountains are accessible only along narrow passes cut by cliff-walled rivers. Some of these rivers cascade hundreds of feet down mountain waterfalls. The bar graph shows the heights of some Upcountry waterfalls.

For 6–10, use the bar graph.

6. What is the range of the data?
   170

7. What is the median waterfall height? Which waterfall is that tall?
   60 ft; Secret Falls

8. What value could be considered an outlier? Explain.
   200; It is separated from the rest of the data.

9. How would the lengths of the bars change if the interval were 10? If it were 25?
   The bars would be longer; the bars would be shorter.

10. Which waterfall is twice as tall as Secret Falls?
    Brasstown Falls

11. Cristina's family took a 7-day camping and hiking trip to the Blue Ridge Mountains. Their trip began on April 28. On what day did the trip end?
    May 4

12. During the trip, Cristina took a 2-hour hike to Secret Falls. If she left for the falls at 11:30 A.M., what time did she arrive there?
    1:30 P.M.

HEIGHTS OF UPCOUNTRY WATERFALLS

Waterfalls: Big Bend, Long Creek, Brasstown, Lower Whitewater, Secret

Height (in feet): 0 20 40 60 80 100 120 140 160 180 200 220

Unit 2 Problem Solving
On Location in South Carolina's Blue Ridge Mountains

Purpose To provide additional practice for concepts and skills from Chapters 5–7

South Carolina Standards Measurement II.B.1 Estimate the distance to objects or places and determine the passage of units of time (minutes, hours, days, week, etc.) it will take to reach them; II.B.3 Determine the amount of elapsed time in hours and minutes within a 12-hour period; Data Analysis and Probability I.C.2 Read and interpret information from tables, line graphs, and bar graphs; I.D.1 Describe types of graphs that may be used to represent categorical data.

Using the Pages

Direct students' attention to Exercise 2. **How does place value help you order the mountain heights from highest to lowest?** Possible answer: I start at the left, at the greatest place value where the digits differ (thousands); then I move right, ordering each place value from greatest to least.

Discuss units of time with students before referring to Exercise 3. **In Exercise 3, how can you change days to hours?** Possible answer: One day is equal to 24 hours, so two days is 24 + 24, or 48 hours.

Clarify the definition of *interval* before referring students to Exercise 9. Then ask: **Why would the graph with an interval of 25 be difficult to read?** Possible answer: It would be difficult because most of the data would not fall on a scale line.

Remind students about the number of days in the different months of the year. Then ask: **What fact do you need to know to solve Exercise 11?** The number of days in April is 30 days.

© Harcourt

Identify the Missing Number

LESSON 9.6A

▶ **Learn**

The Greenville Pavilion Ice Rink is one of two public indoor ice rinks in South Carolina. Each week, 4 pairs of skate laces are replaced. Find the number of laces replaced after 5 weeks.

INPUT	a	1	2	3	4	5
OUTPUT	c	4	8	12	16	■

You can find a rule and then use the rule to find the missing number in the table.

Pattern: Each output is the input multiplied by 4.
Rule: Multiply by 4.
Input: 4   Output: $5 \times 4 = 20$

So, the number of laces replaced after 5 weeks is 20.

EXAMPLES

Find a rule. Name the missing number in the table or pattern.

A

INPUT	x	18	24	30	36
OUTPUT	y	3	4	5	■

Look for a pattern. Test the pattern.

Pattern: Each output is the input divided by 6.

Rule: Divide by 6.

Input: 36   Output: $36 \div 6 = 6$

So, the missing number is 6.

B 1, 2, 4, 8, 16, ■

- Look for a pattern. Possible rule: add 1 or multiply by 2.
- Test the rules for each pair of numbers. Multiply by 2 matches the pattern.
- Use the rule to find the missing number.
  $16 \times 2 = 32$.

So, the missing number in the pattern is 32.

Unit 3 | **South Carolina PACT Success! 9**

▶ **Check**

1. **Explain** the steps you would use to find a missing number in a number pattern.
   Possible answer: Look for a pattern to help you find a rule. Test the rule. Then use the rule to find the missing number.

Find a rule.

2. 3, 9, 27, 81

   multiply by 3

3.

INPUT	r	25	30	35	40
OUTPUT	s	5	6	7	8

   divide by 5

▶ **Practice and Problem Solving**

Find a rule.

4.

INPUT	x	8	9	10	11
OUTPUT	y	32	36	40	44

   multiply by 4

5.

INPUT	w	24	32	40	48
OUTPUT	x	3	4	5	6

   divide by 8

Find a rule. Name the missing number.

6. 400, 200, 100, ■
   divide by 2; 50

7. 45, 54, 63, 72, ■
   add 9; 81

PACT Test Prep

8. This famous pattern is known as Fibonacci's sequence: 1, 1, 2, 3, 5, 8, 13, 21.... Find the rule. Use the rule to find the next number. A I.A.1, I.B.2

   add 2 consecutive terms to get the next number in the pattern; 34

9. Which equation describes the rule for the table in Problem 5?
   A II.B.1, II.C.1

   A $x = w \div 4$ **C** $x = w \times 8$

   B $w = x \times 3$ **(D)** $x = w \div 8$

10. The table at the right shows the cost to rent ice skates at a skating rink. What is the cost of skate rentals for five people? A I.B.2

INPUT	a	1	2	3	4	5
OUTPUT	r	$3	$6	$9	$12	■

   F $10   **(H)** $15

   G $13   **J** $18

10   South Carolina PACT Success! | **Unit 3**

Lesson 9.6A

Objective To find a missing number in a pattern or a table

South Carolina Standards Algebra I.A.1 Create, extend, and analyze numeric patterns (including decimal patterns through thousandths), using models and calculators; I.B.1 Describe and represent number relationships with tables; I.B.2 Determine the rule to identify missing numbers in a sequence or a table; II.B.1 Use variables to represent an unknown quantity using a letter or a symbol; II.C.1 Use equations to represent relationships.

Introduce

WHY LEARN THIS? Finding numbers in a pattern helps you see how one number relates to the previous number.

Suggest that students work with a partner to complete this table. Ask: **What is the rule?** multiply by 6 **Use the rule to find each output number.**

INPUT	m	1	2	3	4	5
OUTPUT	n	6	■	■	■	■
			12	18	24	30

Then ask: **In the pattern 6, 12, 18, 24, 30, ■, what is the next number?** 36 **What is the rule?** add 6

Using the Pages

Ask: **How is finding a missing number in a pattern different from finding a missing number in a table?** Possible answer: In a pattern, you compare each number to the one before it to find and test a possible rule; in an input/output table, you find and test a rule on each pair of numbers.

Assess

Summarize the lesson by asking students to:

Explain how to find the missing number in the pattern 84, 77, 70, 63, 56, ■. What is the missing number? Possible answer: Look for a rule by comparing each number to the previous number and test the rule: subtract 7; 49.

PACT Test Prep

Item	Standard
8	A I.A.1, I.B.2
9	A II.B.1, II.C.1
10	A I.B.2

South Carolina PACT Success!   59

© Harcourt

PROBLEM SOLVING ON LOCATION
in South Carolina's River Basins

South Carolina has over 11,000 miles of rivers and streams. Each of these, from the cold streams in the mountains to the blackwater rivers of the coast, lies within one of the eight river basins that cover the state. The river basins vary in size and shape, and are the areas of land that drain into a lake, river, or stream.

For 1–5, use the table.

1. Which river basins have an area of about 3 million acres?

   Broad and Savannah

2. Could the Broad River Basin's area be 2,545,832? Explain.

   Yes; 2,545,832 rounded to the nearest million is 3,000,000.

3. Write an expression with a variable that can be used to find the combined area of the Saluda River Basin and another basin. What does the variable represent?

   Possible answer: 2,000,0000 + b; b represents the area of the other basin

4. About how many acres do the eight basins cover?

   about 21,000,000 acres

5. Write an expression you could use to find the estimated number of acres in the Santee, Saluda, and Edisto River Basins. Use addition.

   2,000,000 + 2,000,000 + 2,000,000; about 6,000,000 acres

SOUTH CAROLINA'S RIVER BASINS	
River Basin	**Area in Acres (to the nearest million)**
Broad	3,000,000
Catawba-Wateree	2,000,000
Edisto	2,000,000
Pee Dee	5,000,000
Salkehatchie	2,000,000
Saluda	2,000,000
Santee	2,000,000
Savannah	3,000,000

The Edisto River Basin covers over 2 million acres of various types of land. The Edisto River, named by the Native Americans who lived there, is one of the longest free-flowing blackwater rivers in the United States. It flows through about 250 miles of South Carolina's coastal plain.

For 6–8, use the table.

6. Which two types of land cover the same amount of area?

   wetlands, water

7. What if the size of the barren land were twice as large? Write and evaluate the expression.

   9,000 + 9,000; 18,000 acres

8. How many acres does the urban, agricultural, and scrubland areas include? Write an equation with a variable and solve.

   36,000 + 454,000 + 218,000 = n; n = 708,000 acres

9. One weekend, Michael and his family canoed 18 miles down the Edisto River. They canoed the same distance each of the two days. Write an expression that shows how far they canoed each day. Then find the value.

   18 ÷ 2; 9 mi each day

10. In a clean-up project on the Edisto River, each student from a group of 9 collected 3 bags of trash. Write an equation that shows the total number of bags of trash, t, collected by the group. Solve the equation.

    t = 9 × 3; t = 27; 27 bags of trash

EDISTO RIVER BASIN	
Type of Land	**Area in Acres (to the nearest thousand)**
Urban	36,000
Agricultural	454,000
Scrubland	218,000
Barren	9,000
Forest	980,000
Forested wetland	222,000
Wetlands	40,000
Water	40,000

Unit 3 Problem Solving
On Location in South Carolina's River Basins

Purpose To provide additional practice for concepts and skills from Chapters 8–9

South Carolina Standards Number and Operations III.C.1 Round whole numbers to the nearest ten thousand, hundred thousand, and million; Algebra II.B.1 Use variables to represent an unknown quantity using a letter or a symbol; II.C.1 Use equations to represent relationships; Data Analysis and Probability I.C.2 Read and interpret information from tables, line graphs, and bar graphs.

Using the Pages

Direct students' attention to Exercise 5. **How can you use estimation and multiplication to find about how many acres are in the Santee, Saluda, and Edisto River Basins?** Possible answer: about 6,000,000 acres; Since the area of each of the three river basins is about 2,000,000 acres, you can add 3 groups of 2,000,000.

Elicit from students suggestions of words that match expressions. Then ask students to look at Exercise 7. **Identify words in the problem that tell you that the value has to double.** twice as large

Review the definition of the term *equation*. Ask students to give examples of equations. Then have them look at Exercise 8. **How did you write an equation to find the total number of urban, agricultural, and scrubland acres?** Possible answer: I used the table to find the number of acres for each type of land; then I wrote an equation, set it equal to a variable, and solved.

Top worksheet (Lesson 10.1A page)

Name _____

Factor Patterns

▶ **Learn**

At Huntington Beach Park in March, a bird watcher counted 3 nests with 4 baby birds in each nest. In April, he counted 7 nests with 4 baby birds in each nest. How did the total number of baby birds change from March to April?

Huntington Beach Park is one of South Carolina's best areas for watching birds. Terns, gulls, petrels, loons, sparrows, and many other birds nest in this area.

March: $4 \times 3 = 12$

Think: 7 > 3 and 28 > 12

April: $4 \times 7 = 28$

So, the bird watcher counted more birds in April.

MATH IDEA When one factor increases, the product increases. When one factor decreases, the product decreases.

EXAMPLES

A $8 \times 5 = 40$
$5 \times 5 = 25$
So, the product decreased.

B $6 \times 4 = 24$
$6 \times 6 = 36$
So, the product increased.

When one factor is doubled or halved, the product is doubled or halved.

C $3 \times 4 = 12$
$3 \times 8 = 24$
So, the product is doubled.

D $3 \times 10 = 30$
$3 \times 5 = 15$
So, the product is halved.

▶ **Check**

1. **Describe** what happens to the product when one of the factors changes from 4 to 5 and the other factor stays the same. Give an example.
Possible answer: The product increases. 3 × 4 = 12 and 3 × 5 = 15. 15 > 12

Unit 4   South Carolina PACT Success!   13

Write whether the product *increases* or *decreases*.

2. $3 \times 4 = 12$
$3 \times 6 = \blacksquare$
increases

3. $5 \times 9 = 45$
$5 \times 7 = \blacksquare$
decreases

4. $12 \times 5 = 60$
$6 \times 5 = \blacksquare$
decreases

Find the unknown product.

5. $2 \times 8 = 16$
$2 \times 4 = \underline{8}$

6. $6 \times 7 = 42$
$6 \times 9 = \underline{54}$

7. $4 \times 6 = 24$
$8 \times 6 = \underline{48}$

▶ **Practice and Problem Solving**

Write whether the product *increases* or *decreases*.

8. $8 \times 4 = 32$
$8 \times 8 = \blacksquare$
increases

9. $7 \times 3 = 21$
$9 \times 3 = \blacksquare$
increases

10. $10 \times 10 = 100$
$10 \times 1,000 = \blacksquare$
increases

Find the unknown product.

11. $5 \times 6 = 30$
$5 \times 12 = \underline{60}$

12. $7 \times 6 = 42$
$7 \times 3 = \underline{21}$

13. $3 \times 2 = 6$
$3 \times 14 = \underline{42}$

PACT Test Prep

For 14–15, use the table.

14. On which day did Molly spot 2 bird nests that each contained 6 woodpeckers? DAP I.C.2

A Thursday   C Saturday
B Friday   D Sunday

Sightings of Redheaded Woodpeckers	
Day	**Number**
Thursday	9
Friday	6
Saturday	12
Sunday	7

15. What operation is used to find the total number of woodpeckers seen from Thursday through Sunday? DAP I.C.2

F addition   H multiplication
G subtraction   J division

16. Which statement best describes the factor pattern in the equations, $4 \times 5 = 20$ and $4 \times 10 = 40$? NO II.B.1
A Both factors stayed the same, so the product stayed the same.
B Both factors decreased, so the product decreased.
C One factor stayed the same, the other factor doubled, so the product doubled.
D Both factors doubled, so the product doubled.

14   South Carolina PACT Success!   **Unit 4**

Lesson 10.1A

Objective To find the effect on the product when one factor is changed

South Carolina Standards Number and Operations II.B.1 Explain the effect on the product when one of the factors is changed; Data Analysis and Probability I.C.2 Read and interpret information from tables, line graphs, and bar graphs.

Introduce

WHY LEARN THIS? You can use mental math and factor patterns to find how much you will pay, for example, for 5 items at half price: 5 × $4 = $20 or at half price 5 × $2 = $10.

Tell students to extend each pattern. Have them work in pairs to find the solutions.

Increase by 2: 12, 14, 16, ▓, ▓, ▓
18 20 22

Decrease by 5: 40, 35, 30, ▓, ▓, ▓
25 20 15

Increase by doubling: 2, 4, 8, ▓, ▓, ▓
16 32 64

Decrease by halving: 400, 200, 100, ▓, ▓
50 25

Using the Pages

Look at the word problem at the top of the lesson page. Encourage class discussion about how the product changed from March to April. Ask: **Which factor changed and how did it change?** The factor, 3, changed to 7; it increased. **How did it affect the product?** The product increased.

Assess

Summarize the lesson. Ask: **How will the product change? Will it increase or decrease? Explain. 12 × 3 = 36 and 12 × 9 = ▓.** The product will increase; it will be 3 times as great; 108.

How will the product change if one of the factors changes from 12 to 6? The product is halved.

PACT Test Prep

Item	Standard
14	DAP I.C.2
15	DAP I.C.2
16	NO II.B.1

LESSON 11.3A

Closer Estimates

 Learn

The New Charleston Lighthouse at Sullivan's Island, South Carolina, is 163 feet tall. About how many feet will the lighthouse keeper walk in 14 days if he goes up and down the stairs once each day?

The New Charleston Lighthouse is the only lighthouse in the United States that has an elevator.

Estimate: $163 \times 2 \times 14 = 163 \times 28 \to 200 \times 30 = 6,000$

So, the lighthouse keeper will walk about 6,000 feet.

You can find a closer estimate by rounding 163 to 160.

EXACT PRODUCT	ESTIMATE
163	160
× 28	× 30
4,564	4,800

So, a closer estimate is 4,800 feet.

Sometimes when you round factors to numbers that are really close to the factors, you can describe an estimate as a *little more* than or a *little less* than the exact product.

EXAMPLES

A 122×31

EXACT: $122 \times 31 = 3,782$

ESTIMATE: $120 \times 30 = 3,600 \leftarrow$ *a little less than 3,782*

B 279×19

EXACT: $279 \times 19 = 5,301$

ESTIMATE: $280 \times 20 = 5,600 \leftarrow$ *a little more than 5,301*

Another way to estimate is to find two estimates that an exact product is *between*.

EXAMPLE 266×52

LOW ESTIMATE: $200 \times 50 = 10,000$
Round both factors down.

EXACT: $266 \times 52 = 13,832$

HIGH ESTIMATE: $300 \times 60 = 18,000$
Round both factors up.

So, the product 266×52 is *between* 10,000 and 18,000.

▶ **Check**

1. **Explain** two ways to estimate 317×42.
   Possible answer: $300 \times 40 = 12,000$ and $320 \times 40 = 12,800$.

Find an estimate that is closer to the exact product. *Possible answers are given.*

2. $128 \times 43 = 5,504$
   Estimate: $100 \times 40 = 4,000$
   Closer Estimate: $130 \times 40 = 5,200$

3. $25 \times 77 = 1,925$
   Estimate: $30 \times 100 = 3,000$
   Closer Estimate: $30 \times 80 = 2,400$

▶ **Practice and Problem Solving**

Find an estimate that is closer to the exact product. *Possible answers are given.*

4. $238 \times 22 = 5,236$
   Estimate: $200 \times 20 = 4,000$
   Closer Estimate: $240 \times 20 = 4,800$

5. $151 \times 38 = 5,738$
   Estimate: $200 \times 40 = 8,000$
   Closer Estimate: $150 \times 40 = 6,000$

Give two estimates that the exact product is between. *Possible answers are given.*

6. $34 \times 38 = 1,292$
   Low Estimate: $30 \times 30 = 900$
   High Estimate: $40 \times 40 = 1,600$

7. $27 \times 155 = 4,185$
   Low Estimate: $20 \times 100 = 2,000$
   High Estimate: $30 \times 200 = 6,000$

PACT Test Prep

For 8–9, use the table.

Lighthouse	Location	Tower Height	Year Built
Leamington	Hilton Head	94 ft	1880
New Charleston	Sullivan's Island	163 ft	1962

8. The keeper of the New Charleston Lighthouse walked the tower 10 times. Which equation shows an estimate that is a little more than the exact answer? NO III.C.4
   A $200 \times 10 = 2,000$
   B $190 \times 10 = 1,900$
   Ⓒ $170 \times 10 = 1,700$
   D $160 \times 10 = 1,600$

9. Suppose the keeper of the Leamington Lighthouse walked the tower 18 times. What estimate is between the low and high estimate? NO III.C.4
   F 650 ft   H 850 ft
   G 800 ft   Ⓙ $1,800$ ft

Lesson 11.3A

Objective To estimate products that are closer to the exact answer or between a high and a low estimate

South Carolina Standards Number and Operations III.C.4 Refine estimates using terms such as, closer to, between, and a little more than.

Introduce

WHY LEARN THIS? You can make decisions based on an estimate when shopping for food or clothing.

Make sure that students understand the terms *more than* and *between*. Suppose you estimate 168×89 as follows: $200 \times 90 = 18,000$. Ask: **How do you know that the estimate is *more than* the exact answer?** because both factors were rounded up Then ask students to define *between* by asking: **What number is halfway between 400 and 500?** 450 **between 1,000 and 2000?** 1,500 **between 35,000 and 55,000?** 45,000

Using the Pages

Discuss the Example at the bottom of page 15. Ask: **How does this way of estimating help you determine if your answer is reasonable?** Possible answer: If your answer falls between the low and high estimates, it is reasonable.

Assess

Check students' understanding by asking: **How can you estimate 298 × 42 so that the estimate is close to the exact product, 12,516?** Possible answer: Round 298 to 300 and 42 to 40; $300 \times 40 = 12,000$; the estimate is close to the exact product.

Find a high and a low estimate that the exact product is between. Estimate: 458 × 77. $500 \times 80 = 40,000$; $400 \times 70 = 28,000$; The exact product is between 28,000 and 40,000.

PACT Test Prep

Item	Standard
8	NO III.C.4
9	NO III.C.4

© Harcourt

PROBLEM SOLVING ON LOCATION
Along South Carolina's Coastline

LIGHTHOUSES

A lighthouse is a tower that sends a beam of light out to sea. It guides ships to navigate safely along coastlines and waterways. Lighthouses have been built for at least 3,000 years, and they are still used today.

This is the Georgetown Lighthouse located on North Island.

For 1–4, use the table.

1. The keeper rode the elevator in the New Charleston Lighthouse 12 times a day for 15 days. Estimate the total number of elevator trips by rounding one factor up and one factor down. How does the estimate compare to the product? 10×20 or $20 \times 10 = 200$; The estimate is close to the product of 180.

South Carolina's Tallest Lighthouses		
Lighthouse	**Location**	**Height (in ft)**
Cape Romain	McClellanville	150
Georgetown	North Island	87
Hunting Island	Hunting Island State Park	136
Leamington	Hilton Head	94
New Charleston	Sullivan's Island	163
Old Charleston	Morris Island	165

2. Estimate the total number of feet climbed, if the keeper of the Georgetown climbed to the top once a day for 28 days. How would the estimate change if the keeper climbed the Georgetown for 18 days? Explain. $90 \times 30 = 2,700$; about 2,700 ft; The estimate would be less because $90 \times 20 = 1,800$ and $1,800 < 2,700$.

3. Order the heights of the six lighthouses from tallest to shortest. 165, 163, 150, 136, 94, 87

4. The keepers of the Old Charleston and Cape Romain estimated the number of feet climbed during a 30-day period. If each height is rounded up, which estimate is closer to the exact answer? Explain. Old Charleston; the estimate for both is 6,000; $200 \times 30 = 6,000$; 165 is closer to 200.

HUNTING ISLAND STATE PARK

The only lighthouse in South Carolina that is open to the public is the Hunting Island Lighthouse. This lighthouse, built in 1875, replaced the original 1859 structure. Hunting Island State Park was established at the site of the lighthouse in the 1930s, and today is a popular camping and beach destination.

HUNTING ISLAND STATE PARK CAMPING FEES		
	April–October	**November–March**
Campsite	$22/night	about $18/night
	October–April	**May – September**
Cabin	$234 for 3 nights	about $468 for 1 week

For 5–9, use the table.

5. At the park, there are a total of 200 campsites. Find the total fee for one summer night if all sites are rented. $4,400

6. During the busy season, the average fee for renting a cabin for one week is $468. What is the cost for four weeks? $1,872

7. A large group of campers reserved 29 tent sites for one night in November. Estimate the total fee for the group. Tell how you rounded each factor. Possible answer: about $600; I rounded 29 to 30 and 18 to 20, so $30 \times \$20 = \$600$.

8. How much does it cost to rent one campsite for two weeks during December? $252

9. A cabin can sleep 6 to 10 people, while a campsite can accommodate a 2-person tent. For three nights in October, would it cost less for a family of 6 to rent one cabin or three campsites? Explain. three campsites; $3 \times \$22 = \$66$ for three campsites; $\$66 \times 3 = \$198$ for three nights; $\$198 < \$234$

Unit 4 Problem Solving
On Location Along South Carolina's Coastline

Purpose To provide additional practice for concepts and skills from Chapters 10–12

South Carolina Standards Number and Operations II.B.1 Explain the effect on the product when one of the factors is changed; III.C.2 Estimate the product of whole numbers with one factor, 2 digits or less and the other factor, 3 digits or less and determine the reasonableness of the results; III.C.4 Refine estimates using terms such as, closer to, between, and a little more than; Data Analysis and Probability I.C.2 Read and interpret information from tables, line graphs, and bar graphs.

Using the Pages

Invite students to practice estimating products. Have them look at Exercise 1. **How would the product compare to the exact answer if both factors were rounded up?** Possible answer: If 12 were rounded up to 20 and 15 were rounded up to 20, the product would be 400, which is much greater than the exact answer, 180.

Have students look at Exercise 2. **How would the estimated product change if the keeper climbed to the top of the lighthouse for 58 days instead of 28 days?** Possible answer: $90 \times 30 = 2,700$ and $90 \times 60 = 5,400$; so, the estimated product would double.

Explain that estimating is a way to check that an answer is reasonable. Have students look at Exercise 5. **How can you check that your answer is reasonable? Show how you would estimate.** Possible answer: $200 \times 20 = 4,000$; so I know my answer, 4,400, is reasonable.

Exercise 9 could be a challenge to some students. **What steps are needed to solve this problem?** Possible answer: To find the total cost of the 3 campsites, multiply the cost per night by the number of nights and by the number of campsites; $\$22 \times 3 \times 3 = \$198$. Then compare $198 to the cost of a cabin for 3 days. $\$198 < \$234$, so the cost of 3 campsites is less.

LESSON 13.5A

Compare the Quotient and the Dividend

▶ **Learn**

Which has a greater quotient, 36 ÷ 3 or 52 ÷ 3?

Look for a pattern in these division problems.

$\frac{5}{3\overline{)15}}$	$\frac{6}{3\overline{)18}}$	$\frac{7}{3\overline{)21}}$	$\frac{8}{3\overline{)24}}$

As the dividend increases, the quotient increases.

So, since 52 > 36, then 52 ÷ 3 > 36 ÷ 3.

Compare. Look at the divisor.

48 ÷ **4** = 12	The dividend, 48, is 4 times as great as the quotient, 12.
72 ÷ **4** = 18	The dividend, 72, is 4 times as great as the quotient, 18.
96 ÷ **4** = 24	The dividend, 96, is 4 times as great as the quotient, 24.

- In each problem, what does the divisor, 4, tell you about the size of the dividend compared to the quotient? that the dividend is 4 times as great as the quotient

EXAMPLES

A $\frac{12}{5\overline{)60}}$ **B** $\frac{13}{6\overline{)78}}$ **C** $\frac{14}{7\overline{)98}}$

The dividend is 5 times as great as the quotient.

The dividend is 6 times as great as the quotient.

The dividend is 7 times as great as the quotient.

- In 72 ÷ 8 = 9, the dividend is how many times as great as the quotient? 8 times as great

▶ **Check**

1. **Explain** how you can compare the size of the quotient to the dividend in a division problem. Possible answer: The divisor tells you how many times greater the dividend is than the quotient.

Unit 5 South Carolina PACT Success! 19

For 2–4, complete the table.

dividend ÷ divisor	2. 91 ÷ 7	3. 84 ÷ 6	4. 75 ÷ 5
quotient	13	14	15
	The dividend is 7 times as great as the quotient.	The dividend is 6 times as great as the quotient.	The dividend is 5 times as great as the quotient.

▶ **Practice and Problem Solving**

For 5–7, complete the table.

5. 48 ÷ 3 = 16	The dividend, 48, is 3 times as great as the quotient, 16.
6. 68 ÷ 4 = 17	The dividend, 68, is 4 times as great as the quotient, 17.
7. 90 ÷ 5 = 18	The dividend, 90, is 5 times as great as the quotient, 18.

PACT Test Prep

For 8–9, use the table at the right.

8. Ninety-eight fourth graders are visiting the zoo. The teachers divide the students into equal-sized groups so they can spend time at each different exhibit. How many students will be in each group? DAP I.C.2

A 10 **Ⓒ** 14
B 12 **D** 16

9. The students spend 45 minutes at each exhibit and have a 30-minute lunch break. How long are they at the zoo? M II.B.3; DAP I.C.2

F 5 hr **H** 5 hr 30 min
G 5 hr 15 min **Ⓙ** 5 hr 45 min

EXHIBITS AT RIVERBANKS ZOO IN COLUMBIA, SC
African Plains
Aquarium/Reptile Complex
The Birdhouse
Large Mammals
Small Mammals
Sea Lions
Riverbanks Farm

NO II.B.2
10. Some students put 32 photos into 4 albums. Each album has the same number of photos. If the students had 36 photos, would the number of photos in the 4 albums increase or decrease? Explain. Possible answer: Increase; since the number of photos increases (dividend), then the number in each album also increases (quotient).

20 South Carolina PACT Success! Unit 5

Lesson 13.5A

Objective To compare the quotient and the dividend when dividing whole numbers

South Carolina Standards Number and Operations II.B.2 Compare the size of the quotient to the dividend when dividing two whole numbers; Measurement II.B.3 Determine the amount of elapsed time in hours and minutes within a 12-hour period; Data Analysis and Probability I.C.2 Read and interpret information from tables, line graphs, and bar graphs.

Introduce

WHY LEARN THIS? You can look for patterns as you compare the quotient and the dividend in a division problem.

To introduce the lesson, write the following division problem on the board.

$$\begin{array}{r} 5 \\ 7\overline{)35} \\ -35 \\ \hline 0 \end{array}$$

Review these vocabulary words with your students: quotient, divisor, and dividend.

Using the Pages

Discuss Examples A, B, and C. Ask: **Why is it helpful to compare the quotient and the dividend in a division problem?** Possible answer: You can tell what happens to the quotient as the dividend increases.

Direct students' attention to Exercise 10. Explain that the number of photos represents the dividend and the number of albums represents the divisor.

Assess

In 72 ÷ 12 = 6, explain how you can compare the size of the quotient to the dividend. The dividend is 12 times as great as the quotient, 6.

PACT Test Prep

Item	Standard
8	DAP I.C.2
9	M II.B.3; DAP I.C.2
10	NO II.B.2

© Harcourt

Name _____

Change Units of Time

 Learn

Roshanda and her family are driving from Charleston to Clinton. A travel guide says that the trip takes about 3 hours. The computer calculates that the trip will last 180 minutes. Are these amounts of time the same?

Units of Time
60 **seconds (sec)** = 1 minute (min)
60 **minutes (min)** = 1 hour (hr)
24 **hours (hr)** = 1 day

One Way Multiply to change and compare units of time.

3 hours = ■ minutes   To change a larger unit to a smaller unit, multiply.

number of hours	×	number of minutes in 1 hour	=	total minutes
3	×	60	=	180

Another Way Divide to change and compare units of time.

180 min = ■ hr   To change a smaller unit to a larger unit, divide.

number of minutes in 3 hours	÷	number of minutes in 1 hour	=	number of hours
180	÷	60	=	3

So, 3 hours is the same amount of time as 180 minutes.

• Explain how to find the number of seconds in 2 hours.

Possible answer: number of minutes is
2 × 60 = 120 minutes; number of seconds
is 120 × 60 = 7,200 seconds

 Check

1. **Tell** how you would find the number of minutes in 4 hours. Multiply the number of hours by the number of minutes in one hour; 4 × 60; 240 min.

Complete.

2. 5 hr = ■ min   3. 240 sec = ■ min   4. 1 hr = ■ sec
   300   4   3,600

Practice and Problem Solving

Complete.

5. 120 sec = ■ min   6. 360 min = ■ hr   7. 48 hr = ■ days
   2   6   2

8. 2 hr = ■ min   9. 6 days = ■ hr   10. 300 sec = ■ min
   120   144   5

11. Louis and Tim left for the soccer game at 2:20 P.M. They arrived back home at 5:35 P.M. How long were they gone?
    3 hr 15 min

12. Anthony drives 75 minutes to his uncle's home in Myrtle Beach. The return trip takes 63 minutes. How long is the round trip?
    138 min, or 2 hr 18 min

PACT Test Prep

For 13–14, use the line graph.

13. In which month is the normal temperature the greatest? DAP I.C.2
    A Jun   C Aug
    (B) Jul   D Sep

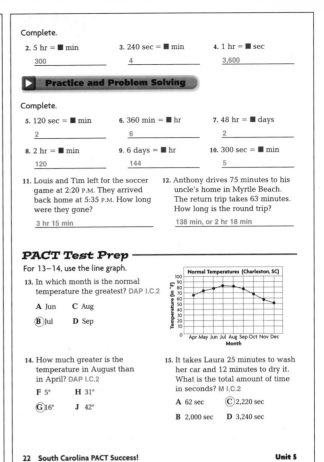

14. How much greater is the temperature in August than in April? DAP I.C.2
    F 5°   H 31°
    (G) 16° J 42°

15. It takes Laura 25 minutes to wash her car and 12 minutes to dry it. What is the total amount of time in seconds? M I.C.2
    A 62 sec   (C) 2,220 sec
    B 2,000 sec   D 3,240 sec

Lesson 15.4A

Objective To change units of time into days, hours, minutes, or seconds

South Carolina Standards Measurement I.C.2 Convert units of time including days, hours, minutes, and seconds; Data Analysis and Probability I.C.2 Read and interpret information from tables, line graphs, and bar graphs.

Introduce

WHY LEARN THIS? You will learn how to solve problems that involve changing units of time to a larger or a smaller unit.

Demonstrate to students how input/output tables can help them change units of time. Have students find the rule in order to name the missing numbers. min to hr: divide by 60, 120 ÷ 60 = 2, or 2 hr; hr to min: multiply by 60, 3 × 60 = 180, or 180 min

MINUTES	60	120	■	180
HOURS	1	■	3	

2

Using the Pages

Students may need more explanation in how to change hours to seconds. Have students use the given example, 3 hr = 180 min = ■ sec. **Explain how to change 3 hours to seconds.** First multiply the number of hours by 60 to change hours to minutes: 3 × 60 = 180; then multiply the number of minutes by 60 to change minutes to seconds: 180 × 60 = 10,800; 10,800 sec.

Assess

Summarize the lesson by asking students to:
Explain how to change 360 seconds to hours.
First, divide the number of seconds by the number of seconds in 1 minute: 360 ÷ 60 = 60; then, to change minutes to hours, divide the number of minutes by the number of minutes in 1 hour: 60 ÷ 60 = 1; 1 hr.

PACT Test Prep

Item	Standard
13	DAP I.C.2
14	DAP I.C.2
15	M I.C.2

© Harcourt

Common Multiples

> VOCABULARY
> common multiple

Learn

South Carolina elects a governor every four years and a U.S. Senator every six years. In 2010 both positions will be up for election. In the next 60 years after 2010, how many times will the election for both positions occur at the same time?

You will need to find common multiples of 4 and 6. A **common multiple** is a multiple shared by two or more numbers.

Multiples of 4: 4, 8, **12**, 16, 20, **24**, 28, 32, **36**, 40, 44, **48**, 52, 56, **60**

Multiples of 6: 6, **12**, 18, **24**, 30, **36**, 42, **48**, 54, **60**

So, the election will occur at the same time 5 different times.

• What are the first three common multiples of 4 and 6? 12, 24, 36

ACTIVITY

• Look at the chart. Shade the numbers that are multiples of 3.

• Draw Xs through the numbers that are multiples of 5.

1	2	3	4	5	6	7	8	9	10
11	12	13	14	15	16	17	18	19	20
21	22	23	24	25	26	27	28	29	30

• Look for the shaded squares that also have an X. Which numbers are common multiples of 3 and 5? 15 and 30

Check

1. **Tell** how to find a common multiple of 2 and 5. Possible answer: Write multiples of each number. Then look for the common multiples.

Find three common multiples for each pair of numbers. Possible answers are given.

2. 3 and 8 24, 48, 72   3. 4 and 12 12, 24, 36   4. 6 and 10 30, 60, 90

Find three common multiples for each pair of numbers. Possible answers are given.

5. 3 and 7 21, 42, 63   6. 8 and 10 40, 80, 120   7. 5 and 7 35, 70, 105

8. 3 and 6 6, 12, 18   9. 1 and 5 5, 10, 15   10. 9 and 12 36, 72, 108

11. Cheryl found some common multiples of 8 and another number. The first three common multiples are 24, 48, and 72. What is the other number? 12 or 24

12. Two hundred eighty-two students are taking a field trip to the state capital. They are divided equally into 6 buses. How many students ride in each bus? 47 students

PACT Test Prep

For 13–16, use the table.

13. If the election for governor and state house rep. occur in 2006, in what year will they both occur again? NO I.G.2; DAP I.C.2

  A 2008   C 2012

  (B) 2010   D 2016

SOUTH CAROLINA ELECTED OFFICIALS	
Office	Term
Governor	4 years
State Senator	4 years
State House Rep.	2 years
U.S. Senate	6 years
U.S. House Rep.	2 years

14. If a South Carolinian is elected to 4 state house rep. and 2 state senate terms, how many years will she serve? DAP I.C.2

  F 6   H 12

  G 8   (J) 16

15. If the state senators and U.S. senators are elected in the same year, how many years later will both elections occur in the same year? NO I.G.2; DAP I.C.2

  A 4   C 8

  B 6   (D) 12

16. Which official will serve the longest period of time—a governor who serves 2 terms, a state senator who serves 3 terms, or a U.S. senator who serves 2 terms? DAP I.C.2

  U.S. senator and state senator; 12 yr = 12 yr

Lesson 16.1A

Objective To find common multiples of pairs of whole numbers

South Carolina Standards Number and Operations I.G.2 Determine common multiples of pairs of whole numbers each of which is less than or equal to 12; Data Analysis and Probability I.C.2 Read and interpret information from tables, line graphs, and bar graphs.

Vocabulary common multiple

Introduce

WHY LEARN THIS? You can determine how many times you will have to do two different chores in the same week.

To introduce common multiples, present this situation to the class:

> Alex and Ben each have a bag of marbles. Alex skip-counts his marbles by twos and Ben skip-counts his marbles by threes.

Ask: **What operation can you use to find the multiples of each number?** multiplication **What is the first number that they both say?** They both say 6.

Using the Pages

Show students how to use a number line to find common multiples. Draw and label a number line from 1 to 25. Have one student draw an X above each number that is a multiple of 4. Have a second student to draw an X below each number that is a multiple of 6. Ask: **How can you identify the common multiples of 4 and 6?** Possible answer: Look for the numbers that have an X above them and below them. **What are two common multiples of 4 and 6?** 12 and 24

Assess

To summarize, ask: **What are three common multiples of 4 and 8?** Possible answer: 8, 16, 24

PACT Test Prep

Item	Standard
13	NO I.G.2; DAP I.C.2
14	DAP I.C.2
15	NO I.G.2; DAP I.C.2
16	DAP I.C.2

Name _____

PROBLEM SOLVING ON LOCATION
at South Carolina's Festivals

Orangeburg

FESTIVAL OF ROSES

The South Carolina Festival of Roses takes place at the Edisto Memorial Gardens in Orangeburg. The 3-day festival includes exhibits, performances, music, and competitions. Some of the tournaments during this festival include softball, golf, basketball, horseshoes, and tennis.

FESTIVAL OF ROSES SCHEDULE			
	Friday	**Saturday**	**Sunday**
People Movers Shuttle	10:00 A.M.–5:00 P.M.	9:00 A.M.–6:00 P.M.	Noon–6:00 P.M.
Arts & Crafts Exhibit	Noon–5:00 P.M.	9:00 A.M.–6:00 P.M.	Noon–6:00 P.M.
Sonshine the Clown		11:00 A.M.–4:00 P.M.	2:00 P.M.–4:00 P.M.
Duck Race		9:00 A.M.–6:00 P.M.	Noon–6:00 P.M.

For 1–3, use the schedule.

1. How many hours is Sonshine the Clown performing in all? How many minutes is this?
   7 hr; 420 min

2. Julie says that the People Movers Shuttle will run for a total of 540 minutes on Saturday. Is she correct? Explain.
   Yes; 540 ÷ 60 = 9; The shuttle runs from 9:00 A.M.–6:00 P.M. on Saturday, which is 9 hours, so Julie is correct.

3. In the duck race, plastic ducks float down a river to a finish line. If a duck race lasts about 12 minutes, how many duck races could take place in one hour? Write an equation and solve.
   Possible answer: $60 ÷ 12 = n$; $n = 5$ races

4. Riverside Drive was closed for the festival from 7 A.M. on Friday until 6:30 P.M. on Sunday. How long was Riverside Drive closed for the festival?
   59 hr 30 min

SPOLETO FESTIVAL USA

The Spoleto Festival USA takes place in Charleston, South Carolina. More than 100 performances of dance, music, and theater can be seen in locations throughout the city. Over 60,000 people attend this 17-day festival each year to see a variety of performances and art exhibits by artists from around the world.

5. The Spoleto Festival USA begins on May 25, 2001. What is the date of the last day of the festival?
   June 10, 2001

6. During the festival, opera productions can be seen in the Gaillard Municipal Auditorium. It is the largest indoor theater in Charleston and can seat 2,734 people. What is the greatest number of people that could attend 3 productions in the auditorium?
   8,202 people

7. Evening jazz concerts during Spoleto can be seen on the campus of the College of Charleston. This college was founded in 1770. How many years ago was the College of Charleston founded?
   Students should subtract 1770 from the current year.

8. At Spoleto, a dance production is scheduled to be in the Emmett Robinson Theater. This theater has 290 seats. A tour group traveling on 4 buses is going to the show. There are 43 people on each bus. How many theater seats will the group need? How many seats will be available for other people?
   172 seats; 118 seats will be available.

9. An orchestra's rehearsal is at 10:35 A.M. and lasts for 1 hour 15 minutes. The orchestra then takes a break before the performance which begins at 1:00 P.M. How much time is there from the end of rehearsal to the start of the performance?
   1 hr 10 min

10. **REASONING** If you know the number of people who attended each of 3 ballet performances, explain how you can find the mean. Find the total number of people at the 3 performances, and divide the sum by 3.

Unit 5 Problem Solving
On Location at South Carolina's Festivals

Purpose To provide additional practice for concepts and skills from Chapters 13–16

South Carolina Standards Number and Operations II.A.1 Explain the meaning of a remainder; III.C.3 Estimate the quotient of whole numbers with a 1-digit divisor, a 2-digit divisor, and multiples of 10 and determine the reasonableness of results; Measurement II.B.3 Determine the amount of elapsed time in hours and minutes within a 12-hour period.

Using the Pages

Review the units of time with students. Ask: **In Exercise 1, how can you use a basic fact and a pattern to help you find the number of minutes the clown performed?** Possible answer: by using the number of hours, 7, and the number of minutes in an hour, 60; $7 × 6 = 42$, so $7 × 60 = 420$

Focus students' attention on Exercise 4. Ask them to use the table and give additional examples of elapsed time. **Have students describe the steps used to solve Exercise 4.** Possible answer: There are 24 hours in one day; from 7 A.M. Friday to 7 A.M. Sunday is 2 days; $24 + 24 = 48$ hours; then count forward from 7 A.M. Sunday to 6:30 P.M., or 11 hours 30 minutes. Then add: 48 hours + 11 hours 30 minutes = 59 hours 30 minutes.

Help students understand that Exercise 8 is a multistep problem. Ask: **How do you know what operations are needed to solve the problem?** Possible answer: To find 4 groups of 43 multiply: $4 × 43 = 172$; to find the difference in the number of seats left, subtract: $290 − 172 = 118$; 118 seats.

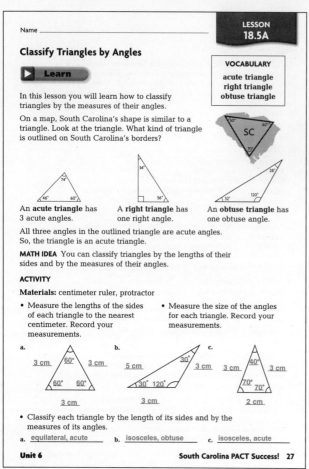

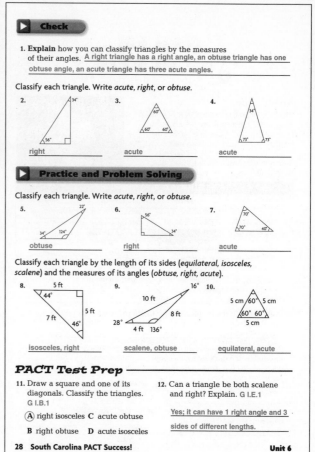

Lesson 18.5A

Objective To classify triangles by the lengths of their sides and the measures of their angles

South Carolina Standards Geometry I.B.1 Classify triangles by lengths of sides (scalene, isosceles, or equilateral) and sizes of angles (acute, obtuse, or right); I.E.1 Make and test conjectures about geometric properties and relationships, and explain their conclusions using models and mathematical vocabulary.

Vocabulary acute triangle, right triangle, obtuse triangle

Materials *For each student* centimeter ruler, protractor

Introduce

WHY LEARN THIS? You will learn how to identify different kinds of triangles used in sculptures, paintings, buildings, and other designs.

Have students look around the room to find examples of shapes that are triangles. Ask: **What are some different ways to classify these triangles?** Possible answers: scalene, isosceles, equilateral

Using the Pages

Have students read the Learn section and the Math Idea. Then guide them through the Activity.

Ask: **How can you classify a triangle by the lengths of its sides and the measures of its angles?** Determine how many sides have the same lengths and if the triangle has 3 acute angles, one right angle, or one obtuse angle.

Have students share their responses to Exercise 12. Then ask: **Can a triangle be both isosceles and obtuse? Explain.** Yes; It can have 2 sides with the same length and 1 obtuse angle.

Assess

Draw a triangle with 2 equal sides and 3 acute angles. You may use a protractor to help. Name the triangle. Check students' drawings; isosceles, acute.

PACT Test Prep

Item	Standard
11	G I.B.1
12	G I.E.1

PROBLEM SOLVING ON LOCATION
at the Columbia Museum of Art

MODERN ARCHITECTURE

The Columbia Museum of Art in Columbia, South Carolina displays paintings and sculptures of the last 600 years from around the world. Its new facility, which opened in 1998, has over 20,000 square feet of gallery space, and was built using many different geometric shapes.

For 1–7, use the photograph of the entrance hall at the Columbia Museum of Art.

1. Describe the relationship between $\overline{DE}$ and $\overline{FG}$.
   They are parallel line segments.

2. Classify triangle ABC. Write *isosceles*, *scalene*, or *equilateral*.
   isosceles

3. Trace $\angle CAB$. Then use a protractor to measure the angle.
   The measure of $\angle CAB$ is 115°.

4. What kind of angle is $\angle ABC$? Write *right*, *acute*, or *obtuse*.
   acute

5. Tell whether triangle ABC has *rotational symmetry*, *line symmetry*, or *both*.
   line symmetry

6. Identify the shapes of Figures 1 and 2. Tell whether they are *congruent*, *similar*, or *neither*.
   rectangles; congruent

7. Classify the figure $BCJH$ in as many ways as possible. Write *quadrilateral*, *parallelogram*, *rhombus*, *rectangle*, *square*, or *trapezoid*.
   quadrilateral, trapezoid

CLASSIC ART

One of the new items at the Columbia Museum of Art is a French porcelain saucer made in 1790. It has a blue and orange enameled surface which displays a variety of Roman classical figures.

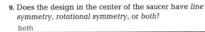

For 8–11, use the photograph of the saucer.

8. Are the different circles on the saucer *congruent*, *similar*, or *neither*? Explain.
   Similar; they have the same shape, but not the same size.

9. Does the design in the center of the saucer have *line symmetry*, *rotational symmetry*, or *both*?
   both

10. Estimate the circumference of the saucer if its radius is 3 inches.
    diameter = 2 × 3 = 6 in.; 6 × 3 = 18; about 18 in.

11. The circumference of the inner circle on the saucer is about 6 inches. What is the diameter of the inner circle?
    6 ÷ 3 = 2; about 2 in.

12. Many of the paintings on display at the museum are on rectangular canvases. Identify the kind of angle that makes up each of the four corners of a rectangular painting.
    right angle

13. A painting at the museum is displayed in a plain rectangular frame. Tell what you know about the opposite sides of the frame.
    The opposite sides are congruent and parallel.

14. A painter uses a canvas that has two pairs of parallel sides, with all four sides the same length. What figures can the canvas be?
    a square or a rhombus

Unit 6 Problem Solving
On Location at the Columbia Museum of Art

Purpose To provide additional practice for concepts and skills from Chapters 17–18

South Carolina Standards Geometry I.B.1 Classify triangles by lengths of sides (scalene, isosceles, or equilateral) and sizes of angles (acute, obtuse, or right); IV.F.1 Connect geometry to other areas of mathematics, other disciplines, and to the world outside the classroom.

Using the Pages

Direct students' attention to Exercise 3. Ask: **In triangle *ABC*, how would you classify the triangle by the measures of its angles?** obtuse

As students discuss the figure in Exercise 7, focus their attention on the parallel line segments. Have students connect $\overline{DF}$ and $\overline{EG}$ to form figure *DEGF*. Ask: **How are quadrilaterals *BCJH* and *DEGF* alike? How are they different?** Alike: Both figures have 4 sides and 4 angles; Different: *DEGF* has 2 pairs of parallel sides, opposite sides congruent, and 4 right angles; *BCJH* has 1 pair of parallel sides, 1 pair of congruent sides, and no right angles.

Refer students to Exercise 10. **Explain how to find the circumference when only the measure of the radius is known.** Possible answer: Since the diameter is twice the radius, you can multiply by 2 to find the diameter; then multiply the diameter by 3 to find the circumference.

For Exercise 14, suggest that students first draw different quadrilaterals to help them solve. **What figures could the canvas be if only opposite sides of the canvas were parallel and congruent?** parallelogram or rectangle

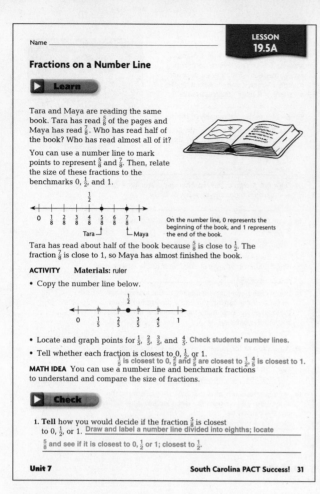

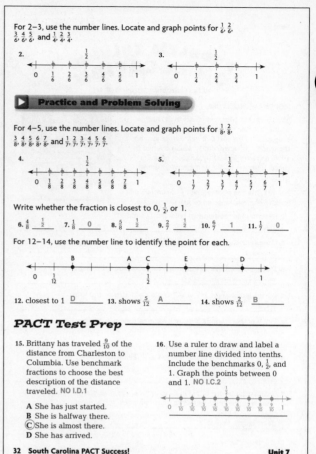

Lesson 19.5A

Objective To locate fractions on a number line and relate the size of fractions to benchmark fractions

South Carolina Standards Number and Operations I.C.2 Locate points on a number line corresponding to a unit fraction and its multiples between 0 and 1; I.D.1 Relate the size of fractions to the benchmark fractions 0, $\frac{1}{2}$, and 1.

Materials ruler

Introduce

WHY LEARN THIS? You can compare fractions when building, cooking, and sewing.

Remind students that they can use a number line to compare and order fractions. Draw the following number line on the chalkboard:

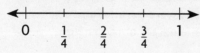

Ask: **On the number line, which is greater, $\frac{3}{4}$ or 1?** 1 **How much greater?** $\frac{1}{4}$ greater

Using the Pages

Discuss the Learn section with your students. Then ask: **Suppose Nick has read $\frac{3}{8}$ of the same book that Tara and Maya are reading. Has he just started the book or has he read about half of the book?** Nick has read about half of the book.

Direct students' attention to the number line above Exercises 12–14. Have them draw the number line on their notebook paper and fill in the missing fractions. Check students' drawings; Ask: **What point shows $\frac{8}{12}$?** E

What point is closest to 1? D

Have students share their drawings for Exercise 16 with the class.

Assess

Tim has traveled $\frac{3}{8}$ of the distance from Florence to Columbia. Use benchmark fractions to describe how far Tim has traveled. Tim is halfway there.

PACT Test Prep

Item	Standard
15	NO I.D.1
16	NO I.C.2

© Harcourt

Compare Decimals

▶ Learn

One of the instruments used to predict weather is a barometer. A barometer measures air pressure. The table shows the air pressures for Columbia, South Carolina and Greenville, South Carolina on May 7, 2001. Which city had the greater air pressure?

AIR PRESSURE
in millimeters of mercury
(May 7, 2001)

City	Pressure
Columbia, SC	30.33
Greenville, SC	30.38

▶ EXAMPLE

Use a place-value chart to compare the decimals.

Think: Line up the decimal points. Compare the digits, beginning with the greatest place value.

Tens	Ones	.	Tenths	Hundredths
3	0	.	3	3
3	0	.	3	8

3 = 3   0 = 0   3 = 3   8 > 3

Since 8 > 3, 30.38 is greater than 30.33.

So, Greenville, South Carolina had the greater air pressure.

MORE EXAMPLES

Compare. Use *greater than, less than,* or *equal to.*

A 2.45 is _?_ 2.36.	B 0.34 is _?_ 0.38.	C 7.80 is _?_ 7.89.
2.45 is *greater than* 2.36.	0.34 is *less than* 0.38.	7.80 is *less than* 7.89.
D 5.18 is _?_ 5.17.	E 3.02 is _?_ 3.02.	F 0.02 is _?_ 0.06.
5.18 is *greater than* 5.17.	3.02 is *equal to* 3.02.	0.02 is *less than* 0.06.

▶ Check

1. **Tell** how to compare the decimals 3.45 and 3.55.

    Possible answer: First compare the digits in each place: ones, 3 = 3; tenths, 5 > 4; so, 3.55 is greater than 3.45.

Compare. Write *greater than, less than,* or *equal to.*

2. 5.7 is _____ 5.8   less than
3. 12.4 is _____ 1.24   greater than
4. 0.05 is _____ 0.15   less than
5. 27.09 is _____ 27.09   equal to

▶ Practice and Problem Solving

Compare. Write *greater than, less than,* or *equal to.*

6. 1.4 is _____ 2.4   less than
7. 0.2 is _____ 0.3   less than
8. 1.24 is _____ 1.14   greater than
9. 1.4 is _____ 1.04   greater than
10. 3.36 is _____ 3.38   less than

PACT Test Prep

11. What are the missing decimals in the following pattern? A I.A.1

    1.76, 2.035, 2.310, ■, ■, 3.135

    A 2.035, 1.76
    (B) 2.585, 2.860
    C 2.585, 3.335
    D 5.06, 7.81

12. Julie pays $186 each month for dance classes. She takes 6 different kinds of dance. To the nearest dollar, is the cost of each dance class *greater than, less than,* or *equal to* $30? NO I.A.2

    (F) greater than
    G equal to
    H less than

13. Glenn's bag weighs 24.75 pounds. Erin's bag weighs 24.25 pounds. Write *is greater than, is less than,* or *is equal to.* NO I.A.2

    The weight of Glenn's bag _____ is greater than

    the weight of Erin's bag.

14. Compare the decimals. Which is true? NO I.A.2

    A 0.12 < 0.02   C 0.73 > 0.78
    (B) 0.45 < 0.54   D 0.89 > 0.99

Lesson 21.5A

Objective To compare decimals by using symbols and words

South Carolina Standards Number and Operations I.A.2 Compare decimals (through hundredths) using symbols (>, <, or =) and words ("is greater than," "is less than," or "equals"); Algebra I.A.1 Create, extend, and analyze numeric patterns (including decimal patterns through thousandths), using models and calculators.

Introduce

WHY LEARN THIS? You will learn how to compare prices of different items at the grocery store.

Remind students of the following symbols and words they can use when comparing decimals.

> means *is greater than*
< means *is less than*
= means *is equal to*

Using the Pages

Guide students through the Examples. Emphasize the importance of lining up the decimal points when comparing decimals.

Do Check Exercises 1–5 with students. Then assign Exercises 6–14.

Direct students' attention to Exercise 8. Ask: **Explain how to use a number line to compare 1.24 and 1.14.** 1.24 is to the right of 1.14 on the number line, so 1.24 is greater than 1.14.

Assess

To summarize the lesson, ask:

Is 30.9 greater than, less than, or equal to 30.19? Explain. less than; Since 1 is less than 9, 30.19 is less than 30.9.

PACT Test Prep

Item	Standard
11	A I.A.1
12	NO I.A.2
13	NO I.A.2
14	NO I.A.2

Name _____

PROBLEM SOLVING ON LOCATION
on South Carolina's Farms

Farmers in South Carolina grow dozens of different crops—some of which are shipped all over the country. One important crop is peaches. The table below shows the approximate fraction of South Carolina's peach trees that are located in each of three general regions.

South Carolina's Peach Trees	
Region	Fraction of Total Trees
Upper State	$\frac{1}{3}$
Ridge	$\frac{1}{2}$
Coastal Plains	$\frac{1}{6}$

For 1–6, use the table.

1. Order the regions from greatest number of trees to least.
Ridge, Upper State, Coastal Plains

2. What fraction of the trees are located in the Ridge and Coastal Plains regions together? $\frac{4}{6}$, or $\frac{2}{3}$

3. What fraction of the trees are located in the Upper State and Coastal Plains regions together? $\frac{3}{6}$, or $\frac{1}{2}$

4. Explain what picture you could draw to represent the fraction of trees in the Coastal Plains region.
Possible answer: a rectangle divided into 6 equal parts, with 1 part shaded

5. How much greater is the fraction of trees in the Upper State region than in the Coastal Plains region? $\frac{1}{6}$ greater

6. Write two equivalent fractions for the fraction of trees in the Upper State region.   Possible answer: $\frac{2}{6}$, $\frac{3}{9}$

Peanuts are another important crop in South Carolina. They are grown in nearly every county in the state. This table shows how many pounds of peanuts were produced in each district in South Carolina in 1999.

1999 South Carolina Peanut Production	
District	Peanut Production (in millions of pounds)
Eastern	1.23
West Central	2.73
Central	11.34
Southern	9.97

For 7–12, use the table.

7. Which district had the greatest peanut production?
Central

8. Did the Eastern and Southern districts combined, produce a greater amount of peanuts than the Central district? Explain.
No; 1.23 + 9.97 = 11.2; 11.2 < 11.34, or 11.2 is less than 11.34

9. How many millions of pounds more did the district with the greatest production produce than the district with the least production?
10.11 million lb

10. Estimate the total production, in millions of pounds, for all four districts.
about 25 million lb

11. Order the four districts from least to greatest production.
Eastern, West Central, Southern, Central

12. What was the production, in millions of pounds, of the Southern district, rounded to the nearest tenth?
10.0 million lb

Unit 7 Problem Solving
On Location on South Carolina's Farms

Purpose To provide additional practice for concepts and skills from Chapters 19–22

South Carolina Standards Number and Operations I.A.2 Compare decimals (through hundredths) using symbols (>, <, or =) and words ("is greater than," "is less than," or "equals"); I.C.1 Describe fractional parts of collections of objects; I.E.1 Write equivalent forms of commonly used fractions; III.D.1 Round decimals to the nearest tenth and hundredth; Data Analysis and Probability I.C.2 Read and interpret information from tables, line graphs, and bar graphs.

Using the Pages

To help students understand adding unlike fractions, discuss Exercise 2. Ask: **What must you do first in order to add these two unlike fractions?** Possible answer: Find equivalent fractions that name the same amount.

Then discuss subtracting unlike fractions. Ask: **In Exercise 5, explain how you subtracted the unlike fractions.** Possible answer: Since I know that two $\frac{1}{6}$ fraction bars are equivalent to $\frac{1}{3}$, I can rename $\frac{1}{3}$ as $\frac{2}{6}$; then subtract: $\frac{2}{6} - \frac{1}{6} = \frac{1}{6}$.

Students can practice comparing, adding, and subtracting decimals in Exercises 8 and 9. Ask: **How much greater is the amount of peanuts produced in the Central district than the Eastern and Southern districts combined?** 0.14 million pounds greater

Refer students to Exercise 10. **How did you estimate the total production of peanuts in all four districts?** Possible answer: Round each decimal to the nearest whole number; then add: 1 + 3 + 11 + 10 = 25; about 25 million lb.

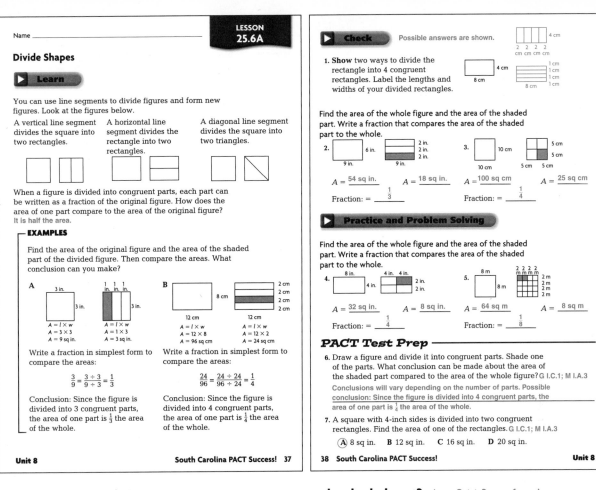

Divide Shapes

▶ Learn

You can use line segments to divide figures and form new figures. Look at the figures below.

A vertical line segment divides the square into two rectangles.

A horizontal line segment divides the rectangle into two rectangles.

A diagonal line segment divides the square into two triangles.

When a figure is divided into congruent parts, each part can be written as a fraction of the original figure. How does the area of one part compare to the area of the original figure? It is half the area.

─ EXAMPLES ─

Find the area of the original figure and the area of the shaded part of the divided figure. Then compare the areas. What conclusion can you make?

A

$A = l \times w$
$A = 3 \times 3$
$A = 9$ sq in.

$A = l \times w$
$A = 1 \times 3$
$A = 3$ sq in.

Write a fraction in simplest form to compare the areas:

$$\frac{3}{9} = \frac{3 \div 3}{9 \div 3} = \frac{1}{3}$$

Conclusion: Since the figure is divided into 3 congruent parts, the area of one part is $\frac{1}{3}$ the area of the whole.

B

$A = l \times w$
$A = 12 \times 8$
$A = 96$ sq cm

$A = l \times w$
$A = 12 \times 2$
$A = 24$ sq cm

Write a fraction in simplest form to compare the areas:

$$\frac{24}{96} = \frac{24 \div 24}{96 \div 24} = \frac{1}{4}$$

Conclusion: Since the figure is divided into 4 congruent parts, the area of one part is $\frac{1}{4}$ the area of the whole.

Unit 8   South Carolina PACT Success! 37

▶ Check   Possible answers are shown.

1. **Show** two ways to divide the rectangle into 4 congruent rectangles. Label the lengths and widths of your divided rectangles.

Find the area of the whole figure and the area of the shaded part. Write a fraction that compares the area of the shaded part to the whole.

2. $A = \underline{54 \text{ sq in.}}$ $A = \underline{18 \text{ sq in.}}$ Fraction: $= \underline{\frac{1}{3}}$

3. $A = \underline{100 \text{ sq cm}}$ $A = \underline{25 \text{ sq cm}}$ Fraction: $= \underline{\frac{1}{4}}$

▶ Practice and Problem Solving

Find the area of the whole figure and the area of the shaded part. Write a fraction that compares the area of the shaded part to the whole.

4. $A = \underline{32 \text{ sq in.}}$ $A = \underline{8 \text{ sq in.}}$ Fraction: $= \underline{\frac{1}{4}}$

5. $A = \underline{64 \text{ sq m}}$ $A = \underline{8 \text{ sq m}}$ Fraction: $= \underline{\frac{1}{8}}$

PACT Test Prep

6. Draw a figure and divide it into congruent parts. Shade one of the parts. What conclusion can be made about the area of the shaded part compared to the area of the whole figure? G I.C.1; M I.A.3
   Conclusions will vary depending on the number of parts. Possible conclusion: Since the figure is divided into 4 congruent parts, the area of one part is $\frac{1}{4}$ the area of the whole.

7. A square with 4-inch sides is divided into two congruent rectangles. Find the area of one of the rectangles. G I.C.1; M I.A.3
   (A) 8 sq in.   **B** 12 sq in.   **C** 16 sq in.   **D** 20 sq in.

38 South Carolina PACT Success!   Unit 8

Lesson 25.6A

Objective To subdivide figures and draw conclusions about area and fractional relationships

South Carolina Standards Geometry I.C.1 Subdivide two-dimensional shapes to form new shapes and draw conclusions about area and fractional relationships; Measurement I.A.3 Find the area of geometric shapes using models.

Introduce

WHY LEARN THIS? You can divide your garden into congruent fractional parts and compare the areas needed for various types of plants.

Elicit from students that congruent figures have the same size and shape. Draw examples on the board of 2 figures that are congruent and 2 figures that are not congruent.

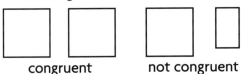

congruent   not congruent

Using the Pages

Discuss Example A. Ask: **Suppose the divided figure has 2 parts shaded. What is the area of**

the shaded part? $A = 2 \times 3$, or 6 sq in.
How much greater is the shaded area now?
Possible answer: It is 2 times as great; $\frac{2}{3} > \frac{1}{3}$. **What new conclusion can you make?** The area of two parts is $\frac{2}{3}$ the area of the whole.

Be sure students understand that the fractional relationship for the area of each subdivided figure is determined by the number of congruent parts. Draw their attention to Exercise 5 where the area is found for 2 shaded parts of the subdivided figure, $A = 4 \times 2$, or 8 sq m.

Assess

Summarize the lesson by asking students: **A rectangle is divided into 5 congruent parts. What fraction compares the area of one part to the whole?** $\frac{1}{5}$ **A rectangle has a length of 6 cm and a width of 2 cm. What is the area of the rectangle?** 12 sq cm **If the rectangle is divided into 4 congruent parts what is the area of each part?**
3 sq cm

PACT Test Prep

Item	Standard
6	G I.C.1; M I.A.3
7	G I.C.1; M I.A.3

South Carolina PACT Success!   73

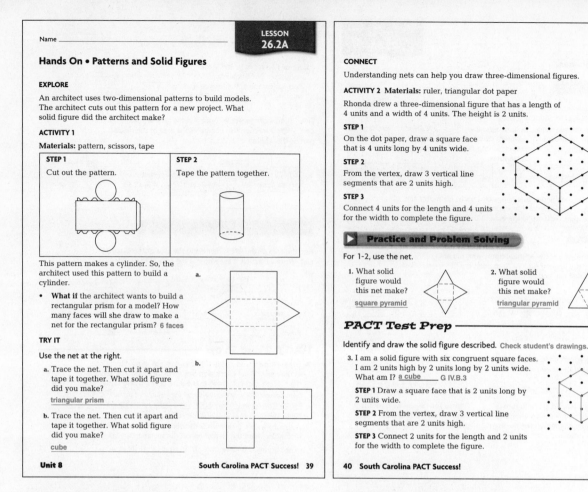

Lesson 26.2A

Objective To build and draw solid figures by their patterns

South Carolina Standards Geometry IV.B.3 Draw two- or three-dimensional objects given a verbal description; IV.C.1 Identify and build rectangular prisms and cylinders from a given two-dimensional representation (net).

Materials *For each student* cylinder pattern, scissors, tape, tracing paper, ruler, triangular dot paper

Introduce

WHY LEARN THIS? You can make models of buildings and other solid objects using a pattern.

Review names of solid figures with students. Have students identify the differences and similarities between a prism and a pyramid and a cylinder and a cone. Make a list on the chalkboard. Then ask: **Where can prisms, pyramids, cylinders, and cones be seen in everyday life?** Possible answer: ice-cream cones, rolls of plastic wrap, cardboard boxes, buildings

Using the Pages

Direct students' attention to the nets in the Try It section. Ask: **How does a net help you build a solid figure?** It allows you to fold the pattern along the fold lines to form the solid figure.

Assess

Check students' understanding by asking: **What solid figure can you make from this net?** a rectangular prism

Explain how you can draw a rectangular prism with a length of 6 units, a width of 3 units and a height of 2 units. Draw a rectangular face that is 6 units long by 3 units wide. Draw 3 line segments from each vertex that are 2 units high. Then connect the line segments.

PACT Test Prep

Item	Standard
3	G IV.B.3

© Harcourt

74   **South Carolina PACT Success!**

PROBLEM SOLVING ON LOCATION
in Myrtle Beach

With more than 90 golf courses, Myrtle Beach and its surrounding area is known as the "Seaside Golf Capital of the World." These golf courses are ranked among the nation's best. One highly acclaimed course is the King's North course at Myrtle Beach National.

1. The second hole at King's North is about 376 yards long. About how many feet is this?

   about 1,200 ft

2. The fifteenth hole, one of the longest on the course, is about 487 yards long. Estimate the length of your step or stride. About how many steps would it take you to walk the 487 yards?

   Answers will vary. Possible answer: about 500–1,000 steps

3. On the twelfth hole, the tournament tee lies 19 yards behind the regular men's tee. Is this greater than or less than 700 inches? Explain.

   Less than; 19 yd = 57 ft; 57 ft = 684 in.; 684 in. < 700 in.

4. The tenth hole is 461 yards long. If a golfer is 150 yards from the hole, how far, in feet, has she already hit the ball?

   933 ft

5. A section of the tee area for the first hole is rectangular shaped and has a total area of 54 square yards. If it is 18 ft wide, how long is it?

   9 yd, or 27 ft

6. The driving distance from Columbia to Myrtle Beach is about 138 miles. Greenville is about 105 miles farther away. About how far is it to drive from Columbia to Greenville?

   about 240 mi

BASEBALL AT MYRTLE BEACH

The Myrtle Beach Pelicans is the city's minor league baseball team. Beginning in the Spring of 1999, the Pelicans' games at Coastal Federal Field have become popular with both local residents and tourists visiting Myrtle Beach.

7. The bases on the field form a square which measures 90 feet on each side. What is the area of the field inside the bases?

   8,100 sq ft

8. During the first game at Coastal Federal Field, a Pelicans player hit a home run in the fourth inning. How far did he run on one trip around the bases?

   360 ft

9. Coastal Federal Field's rectangular scoreboard measures 48 feet tall and 74 feet wide. What is the area of the scoreboard?

   3,552 sq ft

10. Located in left field is a video board that measures 120 inches tall and 156 inches wide. What are these measurements in feet?

    10 ft tall and 13 ft wide

11. The outfield wall is lined with 21 billboards. Each billboard is 24 feet long and has an area of 192 square feet. How tall is each billboard?

    8 ft

12. The field is surrounded by a warning track that is 1,308 feet long. How much shorter is this distance than $\frac{1}{4}$-mile?

    12 ft

13. Baseballs used during the Pelicans' games each weigh about 5 ounces. About how much, in pounds, do 16 baseballs weigh?

    about 5 lb

Unit 8 Problem Solving
On Location in Myrtle Beach

Purpose To provide additional practice for concepts and skills from Chapters 23–26

South Carolina Standards Geometry IV.F.1 Connect geometry to other areas of mathematics, other disciplines, and to the world outside the classroom; Measurement I.C.1 Convert units of measure within the metric system: length (centimeters, meters, kilometers), mass (grams, kilograms); within the customary system: length (inches, feet, yards), weight (ounces, pounds), liquid volume (cups, pints, quarts, gallons); II.B.1 Estimate the distance to objects or places and determine the passage of units of time (minutes, hours, days, week, etc.) it will take to reach them.

Using the Pages

Focus students' attention on the multiple steps involved in Exercise 4. Ask: **What operation do you use to find how far, in yards, the golfer hit the golf ball? Why do you need to change units to solve the problem? What operation do you use to change the units?** subtraction:

461 − 150 = 311 yd; Possible answer: The data is given in yards and the answer should be in feet; multiplication: 311 × 3 = 933; 933 ft.

For Exercise 5, encourage students to draw and label a diagram using the given facts. Ask: **Why do you need to change the given measurement for the width? How does this help you find the length of the unknown side?** Possible answer: The area is given in square yards, so the width should change to yards(18 ÷ 3 = 6, 6 yd); you can divide the area by the width to find the unknown length; 54 ÷ 6 = 9; 9 yd.

Focus students' attention on Exercise 12. Ask: **Describe the steps needed to solve this problem.** Possible answer: To find the number of feet in $\frac{1}{4}$-mile, divide: 5,280 ÷ 4 = 1,320; then subtract: 1,320 − 1,308 = 12, or 12 ft shorter.

Refer students to Exercise 13. Discuss weight in customary units with students. Then ask: **How can you find the weight of 16 baseballs using this information?** Possible answer: Find the weight of 16 baseballs in ounces: 16 × 5 = 80; divide the total number of ounces by the number of ounces in one pound: 80 ÷ 16 = 5; 5 lb.

**LESSON
27.4A**

Multiple-Stage Events

▶ **Learn**

Salita is going to toss this quarter and spin this pointer. What are the possible outcomes? Which outcomes are equally likely?

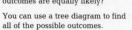

You can use a tree diagram to find all of the possible outcomes.

Since heads and tails are equally likely and blue, red, yellow, and green are equally likely, all of the outcomes are equally likely.

Quarter	Pointer	Outcomes
heads	blue	heads, blue
	red	heads, red
	yellow	heads, yellow
	green	heads, green
tails	blue	tails, blue
	red	tails, red
	yellow	tails, yellow
	green	tails, green

• What is another method you could use to record the possible outcomes of a coin and a spinner? **Possible answer: Make a table or make an organized list.**

EXAMPLES Mike and Alicia are playing a game where they toss a coin and spin the pointer shown at the right. Alicia wins if her outcome is heads on the coin and red on the spinner. Mike wins if his outcome is tails on the coin and blue on the spinner. What are the possible outcomes? Are (heads, red) and (tails, blue) equally likely?

COIN	COLOR		
	Red	**Blue**	**Green**
Heads	heads, red	heads, blue	heads, green
Tails	tails, red	tails, blue	tails, green

Since the red and blue sections are not the same size, red and blue are not equally likely. So, the outcomes (heads, red) and (tails, blue) are not equally likely.

▶ **Check**

1. **Explain** how to find the possible outcomes of tossing a coin and spinning a pointer. **Make an organized list, table, or tree diagram** to find all of the possible outcomes for tossing a coin and spinning a pointer.

For 2–3, use the spinner and coin. Check students' work.

2. Find the possible outcomes of tossing the coin and spinning the pointer. Show your work. heads-blue, tails-blue, heads-green, tails-green

3. Are the outcomes all equally likely? Explain. Yes; each outcome has the same chance of happening.

▶ **Practice and Problem Solving**

For 4–6, use the spinner and coin. Check students' work.

4. Find the possible outcomes of tossing the coin and spinning the pointer. Show your work. heads-red, heads-yellow, tails-red, tails-yellow

5. Look at the possible outcomes. Which outcomes are equally likely? heads-red and tails-red are equally likely; heads-yellow and tails-yellow are also equally likely

6. Which outcomes are more likely than (heads, yellow) and (tails, yellow)? heads-red and tails-red

PACT Test Prep

7. Team A in Mrs. Wright's class made spinners to decide which state symbols to use in their display. Which is a possible outcome? DAP IV.A.1
   (A) tree-fruit   **C** tree-flower
   B bird-fruit   **D** flower-flower

TEAM A

8. Team B used spinners to make posters. Record the possible outcomes of spinning the two pointers. Explain the method used and determine whether outcomes are equally likely. DAP IV.A.1, IV.B.2
palmetto-wren, palmetto-turtle, palmetto-bass, peach-wren, peach-turtle, peach-bass; Check students' methods; each outcome is equally likely.

TEAM B

Lesson 27.4A

Objective To record outcomes of multiple-stage events using tree diagrams and other methods

South Carolina Standards Data Analysis and Probability IV.A.1 Record the outcomes of a multiple-stage event (e.g., tossing two coins), explain the method used, and determine whether they are equally likely; IV.B.2 Construct tree diagrams to list the possible outcomes for multiple-stage events (e.g., tossing 2 coins).

Introduce

WHY LEARN THIS? You can decide what is equally likely to happen when you toss a coin and spin a pointer in an experiment.

Share with students that each year from 1999–2008, quarters for five states will be released. The South Carolina quarter was introduced in 2000. The tails side of the South Carolina quarter shows the state slogan (The Palmetto State), the state bird (Carolina Wren), the state flower (Yellow Jasmine), and the state tree (Palmetto).

Using the Pages

Direct students' attention to the Examples. Ask: **When playing the game, which is the most likely outcome on the spinner if you spin the pointer once? Why?** red; The red section is larger than either the blue or green sections.

Assess

Check students' understanding by asking: **What methods could you use to find the possible outcomes of tossing 2 coins?** Possible answer: Make an organized list, a table, or a tree diagram.

What are the possible outcomes of tossing a coin and spinning the pointer of a spinner with 3 equal sections labeled 1, 2, and 3? heads, 1; heads, 2; heads, 3; tails, 1; tails, 2; tails, 3 **Which outcomes are equally likely?** Each outcome is equally likely.

PACT Test Prep

Item	Standard
7	DAP IV.A.1
8	DAP IV.A.1, IV.B.2

Points on a Grid

▶ **Learn**

The downtown community of Aiken, South Carolina has a pattern of streets laid out in checkerboard style. You can walk on a tour and see many landmarks and historic sites.

ACTIVITY Take your own tour of Aiken to visit some of the sites.

Materials: coordinate grid, markers

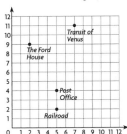

SITE 1 Start at 0. Move right 5 units and then up 2 units. Graph the point and label it "Railroad". Use an ordered pair to describe its location. (5,2)

SITE 2 Start at the Railroad. Then move left 3 units and up 7 units. Graph the point and label it The Ford House. Use an ordered pair to describe its location. (2,9)

SITE 3 Start at the Ford House. Move up 2 units and then right 5 units. Graph the point and label it Transit of Venus. Use an ordered pair to describe its location. (7,11)

SITE 4 Start at Transit of Venus. Move down 7 units and then left 2 units. Graph the point and label it the Post Office. Use an ordered pair to describe its location. (5,4)

▶ **Check**

1. **Describe** the location of another site. If you start at The Ford House and move **right** 8 units and then **down** 6 units you arrive at the Nightingale House. Write the ordered pair.
(10,3)

▶ **Practice and Problem Solving**

In Aiken, there are seven elementary schools. Graph a point and write a label for some of these schools. Use an ordered pair to describe its location. **Check students' graphs.**

2. Start at 0. Millbrook Elementary is located 3 units right and 3 units up.
(3,3)

3. From Millbrook, North Aiken Elementary is located 1 unit right and 4 units up.
(4,7)

4. From North Aiken, East Aiken Elementary is located 2 units right and 2 units down.
(6,5)

5. A class from East Aiken is on a field trip to the historic Joye Cottage Stable. It was originally used as quarters for horses. From East Aiken, the Stable is located 2 units left and 1 unit down. Write the ordered pair.
(4,4)

6. Graph the ordered pairs (1,4), (4,4), (4,1), and (2,3) on a grid. Then label these points A, B, C, and D. Tell how to move point D so that the points connect to form a square. Name the new ordered pair. **Possible answer: Point D can be moved 2 units down and 1 unit left: (1,1).**

PACT Test Prep

For 7–8, use the grid at the right. Square ABCD was moved 6 units down.

7. Describe how point A moved. Then write the ordered pair for its new location. G II.A.1, II.B.2
point A moved 6 units down; (2,4)

8. Which statement describes how point B moved? G II.A.1
A point B moved 5 units over
B point B moved 3 units down
C point B moved 1 unit left and 5 units down
D point B moved 6 units down

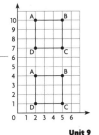

Lesson 30.1A

Objective To identify points on a coordinate grid and describe how to move from one location to another

South Carolina Standards Geometry II.A.1 Describe location and movement using common language and geometric vocabulary, and illustrate with and without technology; II.B.2 Identify and name points on a coordinate grid using an ordered pair of whole numbers.

Materials coordinate grid, markers

Introduce

WHY LEARN THIS? You can use a grid to describe how to get from one place to another at a theme park.

To assess students' understanding of ordered pairs, ask: **How can you use an ordered pair to find a location on a grid?** Possible answer: The numbers in the ordered pair tell you how many units to move to the right of zero and how many units to move up.

Using the Pages

Guide students through the Activity. Point out to students that they will use the ordered pair for Site 1 to find the ordered pairs for Sites 2–4. Stress the importance of checking their work.

After students complete Exercises 7–8, ask: **Look at square ABCD where point D is at (2, 7). Suppose square ABCD was moved 2 units left. Describe how point D moved and write the ordered pair for its new location.** Point D moved 2 units left; (0,7).

Assess

Direct students' attention to the grid on page 45. Have them: **Describe the location of a new site. If you start at the Post Office and move right 5 units and then up 3 units, you arrive at the General Store. Write the ordered pair.** (10,7)

PACT Test Prep

Item	Standard
7	G II.A.1, II.B.2
8	G II.A.1

Name _____

Paths on a Grid

▶ **Learn**

In downtown Columbia, Franco wants to walk from the corner of Main Street and Senate Street to the corner of Sumter Street and Lady Street. He wants to walk the shortest possible path. If he stays on the streets, describe the different paths he can take.

There are three different paths Franco can take. Each path is three blocks long.

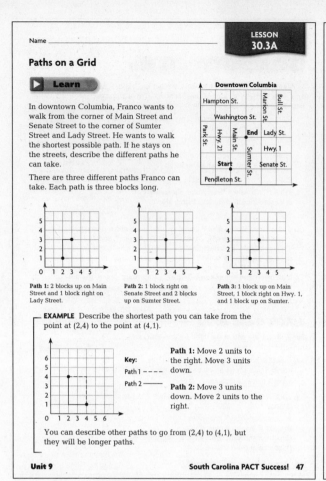

Path 1: 2 blocks up on Main Street and 1 block right on Lady Street.

Path 2: 1 block right on Senate Street and 2 blocks up on Sumter Street.

Path 3: 1 block up on Main Street, 1 block right on Hwy. 1, and 1 block up on Sumter.

EXAMPLE Describe the shortest path you can take from the point at (2,4) to the point at (4,1).

Key:
Path 1 ----
Path 2 ——

Path 1: Move 2 units to the right. Move 3 units down.

Path 2: Move 3 units down. Move 2 units to the right.

You can describe other paths to go from (2,4) to (4,1), but they will be longer paths.

▶ Check

For 1–2, use the map at the right.

1. Describe Franco's path if he walks along Senate Street from Park to Marion.
   Franco walks to the right on Senate St. for 4 blocks.

2. Helena takes the shortest possible path from the corner of Park and Washington to Main and Lady. Describe one possible path.
   Possible answer: Helena walks 2 blocks to the right on Washington and 1 block down on Main.

▶ **Practice and Problem Solving**

For 3–6, use the map above.

3. Tommy walks the shortest path from Sumter and Pendleton to Main and Lady. Describe one possible path.
   Possible answer: 3 blocks up on Sumter, 1 block left on Lady

4. Look at Problem 3. Find the shortest path that Tommy can take if he makes two turns.
   Possible answer: 1 block up on Sumter, 1 block left on Senate, 2 blocks up on Main

PACT Test Prep

5. Greg walks 3 blocks right from the corner of Main and Washington. Where does this path take him? G II.B.1, IV.B.2
   A Marion and Washington
   B Bull and Washington
   C Bull and Hampton
   D Bull and Lady

6. Give directions to tell someone how to get from Bull and Hampton to Main and Hwy. 1 by taking the shortest path. G II.B.1, IV.B.2
   F 3 blocks down, 3 blocks left
   G 1 block down, 3 blocks left, 1 block down
   H 3 blocks down, 4 blocks left
   J 3 blocks left, 4 blocks down

7. Use the grid at the right. Draw two possible paths from (2,5) to (5,2). Describe each path that you drew. G II.B.1, IV.B.2
   Check students' drawings; Possible answer: 3 units down, 3 units right or 3 units right, 3 units down.

Lesson 30.3A

Objective To find possible paths on a grid

South Carolina Standards Geometry II.B.1 Investigate possible paths from one point to another along vertical and horizontal grid-lines; IV.B.2 Describe a path along grid lines from one point to another.

Introduce

WHY LEARN THIS? You can find possible paths from one point to another or identify the shortest path on a map.

Tell students you need to study a map and identify the possible routes when planning a trip. If available, share a map of Columbia with students. Point out that most maps use letters and numbers to identify locations. The numbers usually run horizontally across the map and the letters usually run vertically starting at the top of the map. Remind students that on a grid, a location can be described by first identifying the x-coordinate and then the y-coordinate.

Using the Pages

Look at the Example. Ask: **How does the key show that Path 1 and Path 2 are different?** Possible answer: Starting at (2,4), it shows Path 1 as a dashed line that is right and down and Path 2 as a solid line that is down and right.

Assess

Have students trace the grid used in Exercise 7 on p. 48. Then: **Graph points at (0,1) and (1,2). Describe two possible paths from (0,1) to (1,2).** 1 unit up and 1 unit right or 1 unit right and 1 unit up

Graph points at (2,3) and (4,5). If each path is 4 blocks long and there are no more than 2 turns, how many different paths are possible? 4 **Describe them.** 2 units right and 2 units up; 2 units up and 2 units right; 1 unit right, 2 units up, and 1 unit right; 1 unit up, 2 units right, and 1 unit up

PACT Test Prep

Item	Standard
5	G II.B.1, IV.B.2
6	G II.B.1, IV.B.2
7	G II.B.1, IV.B.2

PROBLEM SOLVING ON LOCATION
in Darlington County

Darlington County is located in the northeastern part of the state. The Darlington Raceway and Kalmia Gardens are among its most popular attractions. Part of Darlington County's appeal is its moderate temperatures. This table shows the average monthly temperatures in Darlington.

For 1–5, use the table.

1. Which month is the coldest? Justify your answer.

   January; it has the lowest average low and high temperatures

2. What is the change in average low temperature from July to December?

   19º

3. Find the median of the average monthly high temperatures.

   25°C

4. The average low temperatures for February to May can be written as the equation $y = 4x - 6$, where x represents the number of the month. For example, February = 2 and March = 3. Graph the equation for the 4 months.

   Check students' graphs.

5. The record low temperature for March is 16 degrees below its average low temperature. Find the record low temperature for March.

   ⁻10°C

AVERAGE TEMPERATURE IN DARLINGTON, SC (IN °C TO THE NEAREST DEGREE)		
Month	Low	High
January	0	13
February	2	15
March	6	20
April	10	25
May	14	29
June	19	32
July	21	33
August	20	32
September	17	30
October	10	25
November	6	20
December	2	15

KALMIA GARDENS AT COKER COLLEGE

Kalmia Gardens is located in Darlington County in Hartsville, South Carolina, and has been open to the public since 1935. A unique 60-foot drop in elevation within the Gardens provides a variety of plant and animal life enjoyed by visitors year-round.

6. Anna says the drop in elevation within Kalmia Gardens is 180 yards. Describe and correct her error.

   Anna multiplied 60 feet by 3. Since 3 feet = 1 yard, she should have divided by 3; 20 yards.

7. Robert lives 8.9 km from Kalmia Gardens. If Melanie lives 5 km less than twice that distance from the Gardens, what is the distance from Melanie's house to Kalmia Gardens? 12.8 km

Jessica planted some of the flowers she had seen at Kalmia Gardens. She used a coordinate grid to map where she planted each type of flower.

For 8–11, use the grid.

8. Write an ordered pair for the location where each flower was planted.

   (1,3) Wisteria; (3,5) Rose; (5,4) Laurel; (8,9) Dogwood

9. Name the plant that is located 2 units down and 2 units left from the rose. Wisteria

10. Name the plant that is located 3 units right and 5 units up from the laurel. Dogwood

11. Name the plant that is located 1 unit down and 2 units right from the rose. Laurel

12. Carl is choosing a project about his visit to Kalmia Gardens. He can make a card, a calendar, a postcard, or a poster. He can feature trees, flowering shrubs, or herbs. In how many ways can he do the project? 12 ways

13. Suppose Jessica wanted to plant a beech tree at (8,10). Describe how you would graph this point on a coordinate grid.

   Start at 0. Move right 8 units. Then move up 10 units. Mark the point.

Unit 9 Problem Solving
On Location in Darlington County

Purpose To provide additional practice for concepts and skills from Chapters 27–30

South Carolina Standards Algebra II.B.1 Use variables to represent an unknown quantity using a letter or a symbol; II.C.1 Use equations to represent relationships; IV.B.1 Describe changes over time as increasing, decreasing, and varying using charts and graphs; Geometry II.A.1 Describe location and movement using common language and geometric vocabulary, and illustrate with and without technology; II.B.1 Investigate possible paths from one point to another along vertical and horizontal grid-lines; II.B.2 Identify and name points on a coordinate grid using an ordered pair of whole numbers; Data Analysis and Probability I.C.2 Read and interpret information from tables, line graphs, and bar graphs.

Materials grid paper

Using the Pages

Direct students' attention to Exercise 4. **After the equation is graphed, what can you observe about the average low temperatures from February to May?** Possible answer: The average low temperatures increased, or got warmer.

Discuss with students how a number line can be used as a thermometer in Exercise 5. Then ask: **Explain how you could you use a number line to help you solve this problem.** Possible answer: You can count back 16° on the number line starting at 6°C; from 6°C to 0°C is a change of 6°; 0°C to ⁻10°C is a change of 10°; so, the record low temperature is ⁻10°C.

Discuss the shortest possible paths on a coordinate grid. Ask students to look at Exercise 11. **Is there another possible path that is the same distance as the one given from the rose to the laurel? Explain.** Yes; move right 2 units and down 1 unit.

Refer students to Exercise 12. Ask: **Why is a tree diagram a good way to find the number of ways Carl can do his project?** Possible answer: A tree diagram shows all the possible outcomes or choices for the project.